T0206333

Innovating Construction Law

Innovating Construction Law: Towards the Digital Age takes a speculative look at current and emerging technologies and examines how legal practice in the construction industry can best engage with the landscape they represent. The book builds the case for a legal approach based on transparency, traceability and collaboration in order to seize the opportunities presented by technologies such as smart contracts, blockchain, artificial intelligence, big data and building information modelling. The benefits these initiatives bring to the construction sector have the potential to provide economic, societal and environmental benefits as well as reducing the incidence of disputes.

The author uses a mixture of black letter law and socio-legal commentary to facilitate the discourse around procurement, law and technology. The sections of the book cover the AS IS position, the TO BE future position as predicted and the STEPS INBETWEEN, which can enable a real change in the industry. The rationale for this approach lies in ensuring that the developments are congruent with the existing frameworks provided by the law. The book proposes various steps that the industry should seriously consider taking from the current position to shape the future of the sector and ultimately create a better, more productive and sustainable construction industry.

This book is a readable and engaging guide for students and practitioners looking to learn more about construction law and its relationship with technology and for those seeking a platform for graduate studies in this area.

Jim Mason is Associate Head of Department, Architecture and Built Environment, Faculty of Environment and Technology, University of the West of England, UK. He is the author of *Construction Law: From Beginner to Practitioner*, published by Routledge.

Innovating Construction Law
Towards the Digital Age

Jim Mason

Routledge
Taylor & Francis Group

LONDON AND NEW YORK

First published 2021
by Routledge
2 Park Square, Milton Park, Abingdon, Oxon OX14 4RN

and by Routledge
52 Vanderbilt Avenue, New York, NY 10017

Routledge is an imprint of the Taylor & Francis Group, an informa business

British Library Cataloguing-in-Publication Data
A catalogue record for this book is available from the British Library

Library of Congress Cataloging-in-Publication Data
Names: Mason, Jim (James Robert), author.
Title: Innovating construction law: towards the digital age / Jim Mason.
Description: Milton Park, Abingdon, Oxon; New York, NY: Routledge, 2021. |
    Includes bibliographical references and index.
Identifiers: LCCN 2020037268 (print) | LCCN 2020037269 (ebook)
Subjects: LCSH: Building laws. | Construction industry—Law and legislation. |
    Construction contracts.
Classification: LCC K891.B8 M375 2021 (print) | LCC K891.B8 (ebook) |
    DDC 343.04/8624—dc23
LC record available at https://lccn.loc.gov/2020037268
    LC ebook record available at https://lccn.loc.gov/2020037269

ISBN: 978-0-367-44349-8 (hbk)
ISBN: 978-0-367-44352-8 (pbk)
ISBN: 978-1-003-00924-5 (ebk)

Typeset in Goudy
by KnowledgeWorks Global Ltd.

To my father Michael John Mason (1935–2007)
and his passion for science fiction

Contents

Figures

Tables

Preface

Law is necessary in order for us to regulate our interactions with one another. The role of the law as the basic civilising force in society was expounded in my first book, *Construction Law from Beginner to Practitioner* (Book One). This new work you are reading is a companion to Book One and reference should be made to it for the background law and an introduction to the construction industry as a consumer of construction law. This book is intended to stand alone from Book One. However, the starting point for this work assumes more in the way of prior knowledge of construction law and reference should be made to Book One as required.

Those wanting to study and practice construction law in particular are seldom disappointed with their chosen field. Construction law covers a broad spectrum of subjects from law involving planning, property, environmental, contract and tort, and can even extend to the law of trusts. It is contract and tort law that provides the main framework in which construction obligations operate. The specialisms, which can then be attached to this basic knowledge, are numerous and supports the notion that there is something for everyone in the broad church of construction law practice. A glance at the reported cases indexes reveals that construction law continues to be a leading area where new principles and precedent are created in the fascinating evolution of the common law. Hadley v Baxendale, still the seminal case on the measure of contractual damages, was described by one judge, as being *"a fixed star in the firmament of precedent."* It is itself a construction case about replacing a mill shaft.[1] Anns v Merton LBC,[2] a case about a poorly installed floor, marked the high water mark of the economic loss doctrine in the law of tort. In more recent times, charting the progress of Building Information Modelling in the case of Trant v Mott Macdonald,[3] which involved access to common data environments, reminds us of the adaptability of the common law to build principles and deal with hitherto unbeknown legal conundrums.

However, the common law's ability to cope is coming under huge strain in the context of the digital and data revolution being experienced. In this book, I wish to explore the view that construction law appears at risk of stagnation in its development generally and more specifically is missing out on the opportunity to be proactive in the face of change in the current digital age. This should not be the case and this work aims to encourage ways in which the perceived gap could be closed. The importance of the law moving with the times was highlighted in 2019 in the LawTech Delivery Panel's Legal Statement on Cryptoassets and smart contracts.[4]

In the author's opinion, construction law writing has a tendency to rake over the same ground and cover only minor changes in well-worn commentaries on the "usual suspects" of dispute generating grey areas. Many readers will be familiar with topics covered in Book One, such as variations, consequential loss, contemporaneous delay analysis and adjudication practice. This perpetuation and introspection of "disputomania" is not helpful in building towards a more collaborative and forward-looking construction industry. There are wider societal responsibilities owed by professionals even when they act entirely at their client's direction. The clients and funders should question whether short-term point scoring in winning a dispute is in their wider interest. I recognise this paradox in my thinking as a solicitor practising construction law. There I was, all set to help clients with my specialist skill and yet with a vested interested in the disputes becoming ever more intractable and my advice about the risks being ignored as clients strengthened their belief in their own unrealistic positions. I discovered plenty of disputes (contentious work) and conflicting agendas in contract drafting (non-contentious work). Clients and contractors, sometimes with rogue tendencies, were seemingly happy to feed the engine of disputes as I settled for the mechanic's role. My observation on this stage of my career has been that the non-contentious work was often more contentious than the contentious. This resonates with the observation that professionals are at least complicit in promoting adversarial attitudes.

Construction lawyers ought, in my opinion, to go to greater lengths to avoid litigation. The responsibility to which I refer was summed up by Abraham Lincoln in the following quote:

> Discourage litigation. Persuade your neighbors (sic) to compromise whenever you can. Point out to them how the nominal winner is often the real loser — in fees, and expenses, and waste of time. As a peacemaker, the lawyer has a superior opportunity of being a good man. There will still be business enough.[5]

The central purpose of this book is to provide a resource for students interested in construction law, and the built environment more widely. I hope that there is something in here for those who seek to strive to make a difference by embracing technology and reducing, and ultimately eliminating, disputes and their accompanying waste. There are two ways to achieve this – collaboration and technological advancement. The former has been around for long enough to demonstrate that it cannot succeed alone. The support of technological advancement is needed firstly by streamlining procedures and then innovating new ways of working. Foremost in the innovations that will achieve this goal are smart contracts – executable computer code delivering transactions in the same way as conventional agreements. In this way, the best example can be set in terms of how a post-industrial society can operate as sustainably as possible within the constraints and with the help of the technology available. Lawyers should seek to take their clients with them down a road of technological enlightenment and thereby earn a level of commendation that they would not otherwise attain.

The partnering or alliancing approach has held sway amongst reformers in the construction industry since its inception through Sir John Egan's Rethinking Construction report.[6] These efforts have continued, largely through the tireless work of Professor David Mosey at Kings College London, where they are being enhanced by the technological developments where collaboration is very much "baked in" to the process. However, a fork in the road may be approaching between a transactional approach and a collaborative approach.

At one level, this book aspires to allow others to take stock of the developments in technology and construction law in a holistic manner as well as providing a starting point for ushering in future developments. There are several different yet potentially complimentary technologies examined and one is not necessarily promoted above the others. It is hoped that the reader will enjoy the variety of approaches to digital aspects, as a singular approach would be to promote one version of the future above the others.

References to the Laws of England also refer to the Laws of England and Wales. References to "he" and "him" should be taken to apply equally to "she" and "her." The numbering system used by the book is intended to assist navigation around the various chapters, and there are cross-references made in the text.

Jim Mason,
Hawkswood College May 2019

Notes

1. Hadley v Baxendale [1854] EWHC Exch J70
2. Anns v Merton LBC [1978] AC 728
3. Trant v Mott Macdonald [2017] EWHC 2061
4. Available at: https://technation.io/about-us/lawtech-panel [accessed on 11 May 2020]
5. *The Collected Works of Abraham Lincoln*, edited by Roy P. Basler, Volume II, "Notes for a Law Lecture" (July 1, 1850), p. 81
6. Egan, J. (1998) *Rethinking Construction – The Report of the Construction Taskforce*, Department of the Environment, Transport and the Regions

Acknowledgements

Thanks to my colleagues from the Department of Architecture and Built Environment at the University of the West of England (UWE Bristol). Nearly twenty years into my academic career, I continue to enjoy the interaction with the variety of professionals and academics in our grouping and the frequent ribbing and collaboration that occurs. It is always intriguing how the different approaches lead to consensus via some healthy debates. Further thanks to the UWE Bristol Estates team, in particular Helen Baker and Mike Ford for sharing their best practice and ideas with us.

Thanks to my mother, Mrs Edith Walsh for the paintings, which grace the pages of this work, and to Jake Mason, my elder son, for working on the cover design also for my younger son, Patrick Mason, for being a model student in peculiar times.

Thanks to the alumni from the Masters in International Construction Law at UWE Bristol – your desire to learn has helped me search for some answers and connections, which I hope come out of these pages. Special mention to Nigel Aqui, current PhD student, for his contribution to the Chapter on Online Dispute Resolution, and to Alan Mcnamara, PhD student in Sydney, Australia, whom I wish every luck with his smart contract prototypes. Thank you to Nana Manu-Mabiza for proof-reading and joining us at UWE Bristol to strengthen our construction law specialism.

I acknowledge the positive influence and support of the Society of Construction Law and thank the following for their hard work, clear thought and progressive attitudes in the field David Mosey, Antony Lavers, Tony Bingham, Rudi Klein and Shy Jackson.

Glossary of terms used relating to technology

This glossary features my definitions of the technology-relevant terms found in the book within the context they are used. The examples cited are taken from the construction field. Computer science will doubtless provide more nuanced and sophisticated definitions of the concepts and I defer to that body of work. Students of technology subjects can identify portal concepts that are the gateways to a wider understanding and appreciation. Again, only some of the more easily accessible portals are covered here. Any gaps or errors in the glossary are due to the author's limitations. A further caveat is to acknowledge that this is a fast-moving field of practice with constant evolution of ideas and concepts. A last caveat is to recognise that spreading oneself too thinly across multiple disciplines may end up with too many superficialities.

Non-lawyers may reasonably ask, "where is the legal glossary?" and I would respectfully refer them to Book One; *Construction Law: From Beginner to Practitioner*.

Artificial intelligence (sometimes neural networks) The logical extension of machine learning can be seen as the ability of a computer to think for itself. This is a huge topic and exercises a good deal of debate at many levels not least of which is philosophical. The position taken in this work is that the human mind is the only place where original thought can occur based on billions of cognitive interactions taking place at the same time. Computers are faster at "computing" but are restricted by their electronic make-up and circuitry from thinking in parallel like their human counterparts. Recent developments in quantum computing may narrow this gap significantly.

Augmented reality This involves the overlaying of a planned development onto the real space/context in which it will operate. For example, a system of pipework can be envisaged in situ in the part built shell of a building. A BIM application can be accessed through the Augmented Reality headset and the pre- and post-installation of the pipework checked to avoid any errors or lack of space for the pipe runs. The consequent savings in error reduction and simple quality control by exception checking are major benefits. Remembering which reality one is working in can become a challenge!

Big data The results of internet of things reporting to cloud-storage systems. The key challenge here is to handle and make sense of the volume of data generated.

BIM Building information modelling (BIM) or digital design and construction represents the most familiar technological advancement within the built environment in the first twenty years of the 21st century. The central idea is that an asset is built first in a virtual world and perfected there before building it in the physical world. BIM is viewed in this work as a portal technology and an enabler of future developments.

Blockchain/Distributed ledger technology This is a digital platform for the recording of transactions or events or facts, which is based on a distributed ledger. Cloud computing means that access to the same list, or chain of data can be shared instantaneously with multiple users from a small handful to hundreds of thousands. The transparent and shared nature of the chain and the fact that all the permitted users can verify that the common version is correct, reduces the scope for illegality or double recording of transactions. This is also known as the immutable nature of the blockchain. Adding new data to the chain requires the consensus of participants. This is known as a consensus algorithm. This is a platform that could support the use of smart contracts. There is a distinction between online and offline ledgers and public and private networks.

Cloud storage Increasingly, databases are not stored on one computer's hard-drive but are shared in an external storage system, the capabilities and size of which appear to have grown exponentially in recent years. These external storage systems are known as cloud computing. This facility is necessary given the huge volume of data generated through such applications as the internet of things and blockchain.

Crypto-currency This is a form of currency which is not reliant on intermediaries such as banks and/or state-run centralised financial controls. Bitcoin is the most recognisable name in this field. Crypto-currencies are known for their volatility in their trading price when compared to established currencies.

Data This is information in the form of facts or opinion based on user experience. Meta-data can be described as data about data, which lends itself to data analytics.

Data analytics The "funnelling" of data into a manageable format from which machine leaning and other uses can occur. Sometimes it is represented in a dashboard of indicators from which decisions can be made and performance monitored.

Data mining This represents the use of computers to approve and verify public blockchain transactions. One approver is chosen at random to receive a reward paid in crypto-currency for performing the verification. Data mining can be performed on a small scale by home computer enthusiasts, right up to on an industrial scale in massive warehouses of computers.

Digital twin This represents the extension of BIM into a digital representation of the building in real time during its operation and maintenance. Digital twins can also exist in a variety of other industries including infrastructure, power networks, highways, river networks and train lines. The vision of Digital Build Britain is to connect these resources together and, eventually, for the twins to operate autonomously based on machine learning and artificial intelligence. This vision can be described as a unicorn.

Drone or UAV An unmanned aerial vehicle, often used in the built environment for surveillance or auditing of work performed and the status of existing infrastructure. For example, a drone can be employed to spot breaches in an electricity supply or to check on the deterioration of a façade to a building.

Exception handling Computer science term for when an external source is required to resolve a discrete point. The external source can be termed an oracle. This might be to resolve a contentious point through online dispute resolution on the meaning of "reasonable endeavours" and other discretionary legal terminology.

Immutability One of the key features of the blockchain is its ability to establish a trustless system. The records entered onto the ledger are immutable in the sense that once they are entered, they cannot be changed without causing anomalies that the other members of the digital community would quickly be alerted towards.

Internet of things To date, the world wide web has been primarily about providing an internet of information. Increasingly, it is the capacity to become an internet of things, or assets, which is being developed. In its simplest form, this consists of data sensors reporting on their status to a mainframe or database where this information is stored. More advanced forms allow for the manipulation of that data for smart processes, for example a digital twin.

Oracle Term used to describe when real-world performance and checks are needed to ensure compliance and reflection of "as planned" performance. The oracle function can be provided remotely or robotically or by a human agent. For example, the completion of an elevation of bricks built by flying robots is completed by the addition of a capstone, the position of which is measured by a hand-held laser-level checker and the work signed off. The need for human input here may be on a specific term of reference only.

Portal technology A "game changer" in the sense that although technology may be superseded by later development, it nevertheless represents a key breakthrough as an enabling standard. For example, BIM and augmented reality.

Provenance The ability to check on the origins of a supplied component. Where something has come from, and the carbon footprint it has created, is becoming increasingly important in a climate changing conscious world. The technologies being described here allow for a much greater degree of traceability and granularity in assessing the provenance of built environment components, from steel to glass to cement.

Robot A machine able to replicate certain human movements and functions automatically.

Smart contracts Computer programming obeys the same logic as law-making in terms of following an instruction. This, together with conditionality, means that the stakeholders on a construction project can trust in the code to plan, execute, verify and pay for the completion of activities. A distinction can be drawn between smart contracts which merely perform obligations inside a "normal" contract and a smart contract which represents the formation as well as the execution of its terms. The latter is a greater challenge to our existing legal frameworks.

Smart technology Smart technology is evidenced by devices and computers either exhibiting machine learning or performing a series of "if this then this" commands (conditionality) with or without the interaction of a human intermediary or certifier. This conditionality is represented in computer science by an algorithm – a process or set of rules to be followed.

Social value The recognition of the importance of taking a wider view of the benefits of the built environment beyond their status or financial viability or opportunity. Increasingly, this involves a consideration of societal and user well-being and the sustainable features of any development. Procurement decisions in the public sector take increasing note of this factor.

Stigmergic This is a concept known in management and societal study. It represents the inter-connectivity of entities in an over-arching attainment of a goal. For example, termites may be said to be behaving stigmergically in the fulfilment of their individual tasks with only a vague sense, if any, of the greater societal good to which they are linked. Stigmergic networks are seen as an improvement on a hub-and-spoke model requiring each actor to report to a central entity.

Unicorn A desirable yet fanciful notion. No one has ever, as far as is provable, seen a unicorn despite wishing it was otherwise. The search for a unicorn is viewed as a worthwhile pursuit in itself as any progress made towards the desired outcome represents an achievement.

Virtual reality A virtual representation of how something will look once it is built. This is mainly used in the built environment at the design stage for client visualisations. Virtual reality is usually facilitated through the use of headsets or other immersive experiences.

Section I
Background

1   Introduction

In her 2017 paper, Margaret Radin[1] wrote about the Ancient Greek Philosopher, Ptolemy (85 to 165 AD), and his geocentric theory which was to dominate how we thought about our place in the universe for 1400 years. Ptolemy spent many years studying the movement of the stars and planets around the Earth and was of the very firm view, and convinced others, that the Earth was the centre of the universe (the geocentric theory). There was one flaw in his grand plan – occasionally, the planets would not follow a regular orbit pattern and go off, on what a judge once deemed, "a frolic of their own."[2] That is, the planets did not follow the orbital pattern.

Ptolemy hit upon an explanation for this inconvenient fact, which is known as Ptolemaic epicycles. This accounted for the variations in speed and direction of the moon, sun and planets. The inconsistencies in the planetary movement were themselves modelled on separate sub-diagrams which sought to explain away the concerns. And yet, the inconvenient fact remained that the epicycles did not make much sense when held up to the geocentric theory itself. Some 1400 years later, Nicolaus Copernicus proposed a heliocentric system; that is, the planets orbit not around the Earth but around the Sun. Ptolemaic epicycles (Figure 1.1) were exposed as folly – an example of where theories are stretched and manipulated to fit the facts which are apparent.

And so, to the point, contract law (and, hence, construction law) is in the same position as the geocentric theory. We have 18th-century principles being manipulated and mis-shaped to fit into the 21st-century digital legal age. Lecturers teach contract law designed for an age of mill shafts (Hadley v Baxendale[3]) and peculiar cold remedies such as carbolic smoke balls.[4] The argument developed in this work is that the approaches in these theories are no longer fit for purpose and we need to make a similar shift in thinking as the heliocentric system. The proposal is therefore that we move away from contract centred thinking – the 100-page wet-signed contract left in the drawer – to contracts legible by computer programmers and lawyers centred on instantaneous self-execution and payment releasing smart contracts.

Radin gives examples of how contract law is no longer coping including the ubiquitous "I agree" click when ordering online. She points out that the instruction is most likely to say that the user affirms that they have read the terms and agree to them. This common procedure has made liars of us all. Empirical

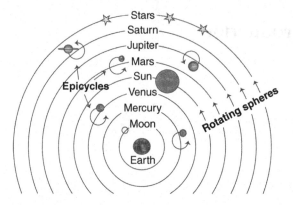

Figure 1.1 Ptomelaic epicycles

research shows that we normally do not read it, and how could we? There is so much of it. Anyway why should we? We cannot change it, and we are not likely to understand it. The objective theory of intent has been stretched and distorted to treat these clicks metaphorically, as if they amount to "acceptance" as if such clicks were the same as saying "I accept" in response to terms that have been actually communicated and understood, and perhaps subject to one's own input, as in bespoke contracts. Routine use of "I agree" is therefore used as an example of the deformation of contract theory that is analogous to a Ptolemaic epicycle.

The Radin paper concludes by saying that contracts have failed to adapt to a set of technological and social conditions that are increasingly far removed from its origins. The counter-view is not being discounted without due consideration. It is true that the common law copes very well with changing times and its main strength is its versatility and ability to create precedent. In the author's view, though, the efforts made to fit the old law to the new era bear some of the characteristics of Ptolemaic epicycles. One "epicycle" is of particular interest in this book as it has dominated construction law best practice for over 30 years – it is the collaborative movement sometimes known as partnering or alliancing.

The term "smart" is in wide popular usage at present and the construction industry is no exception. It seems that every conference and article uses the term in some way, shape or form, as we enter the start of the third decade of the 21st century.

What is actually meant by the term is open to conjecture. A consensus would appear to be that the term denotes some sort of artificial intelligence ultimately replacing human endeavour or making tasks easier to complete. Our attitudes towards these innovations reflect our standing point on matters technological.

Smart contracts and blockchain technology are at the forefront of technological advancement in the financial services industry. The basic premise involves the creation of an automated contract capable of satisfying common conditions and reducing the need for intermediaries in the process. The blockchain is a means by which the transactions can be recorded on a distributed ledger.

Technological progression is much slower in construction than in other industries as evidenced by the slow take-up of UK government-backed initiatives such as the introduction of Building Information Modelling (BIM) and collaborative working agendas. The inevitable trough of disillusionment in relation to smart contracts will centre on concerns about the length of time this will take to achieve and the complexities involved. Every construction project is different, with a specific design and scope of works; this renders contract drafting complex as parties try to account for all possible contingencies.

In this book, the terms "digital technologies" and "data driven technologies" are used interchangeably. It is worth examining here what the terms actually encompass. A helpful approach was summed up neatly in the following quote: *"Digital should be seen less as a thing and more a way of doing things."*[5] The key drivers include creating value at new frontiers typically using sensors and analytics to improve efficiencies of supply-chain operations. Data-driven technologies make more explicit use of the common currency at the heart of both definitions – the data itself. This is the raw material that once harvested can be imagined and repurposed for an amazingly wide range of functions.

Digital technologies represent the advancements we see in automation, proactive decision-making and contextual interactivity. The technologies allow enterprises to be agile, fast, have a much more complete picture of their business and, in this case, the built environment assets they procure, build and maintain.

The value in the digital/data approach is apparent in the usefulness of having, at the author's place of work, a digital campus. An interactive dashboard can bring together all the functionalities from the estate including performance, monitoring and completion of building work and maintenance as they occur. The key here is to access the right level of useful information by analysing and presenting only those indicators useful to the business' search to add value. This approach resonates with the mission of the Centre for Digital Build Britain.

This book is hopefully useful to provide the arguments to persuade the reader to support a movement towards digital technology and to regard this as a necessary and positive force for change within the construction industry. Richard Saxon said that what partnering needed to succeed was BIM.[6] The view postulated here is that what BIM and partnering need to succeed is smart contracts. After all, to borrow the quote from the Saxon Report:

> *Civilisation advances by extending the number of important operations we can perform without thinking about them.*[7]

A book advocating the adoption of digital and data-driven technologies would be a worthwhile undertaking. However, others are better placed to write such guides. The rationale for this book focuses on the connection between construction law and the technologies encountered. Making the case for adoption is only ever part of the equation when discussing progress. Providing the legal mechanisms to either mandate or facilitate change is the greater part of it.

The aim is therefore to draw threads together and lay foundations for knowledge within the context of legal parameters relating to this rarefied area of fast developing practice. However, one has to allow for peaks and troughs in advancement. Take, for example, the development of the smart phone, which appears now to have settled on a pattern of adding more cameras rather than developing any further ground-breaking applications. The halcyon days of the smart phone appear to be over whilst it waits for re-invention. The smart watch, on the other hand (or wrist) looks set to take up the challenge and already provides user health information and monitoring.

Examples of technology moving at a startlingly pace are not hard to find. For example, in March 2016 Google's Deep Mind technology designed a computer program called AlphaGo to play the board game Go. Go is considered much more difficult to win than chess.[8] AlphaGo beat the world's best player in a five-game series. This was a remarkable achievement, in part because commentators thought we were at least a decade away from such a result. The match was also significant for another reason. It is often said, when talking about the limits of machines, that jobs require creativity are safe from automation because creativity is a capacity that machines cannot exhibit. This is what Alan Turing, the Godfather of Computer Science, called the "Lady Lovelace objection."[9] Lady Ada Lovelace argued that a machine could never "originate" or do anything new and could only do whatever we know how to order it to perform. And yet, one particular move by AlphaGo led to exclamations from commentators and was described as "beautiful" by a past Go champion. Contrary to widespread belief, the inference is that machines are now capable of generating novel outcomes, entirely beyond the contemplation of their original human designers.

A similar movement towards change is discernible in automation in construction. This momentum should be seized upon and smart contracts recognised as the most important development to move us from the "as is" position to the "to be" position via the steps in-between. If handled correctly and with the involvement of people in the right mind-set to use them, then a high level of potential benefits can be enjoyed by the industry.

1.1 The legal approaches in the book

The legal approach used in this book follows two traditions that are the black letter approach and the socio-legal. Other approaches exist. However, these have been chosen as they provide a helpful counter-balance to each other. Black letter law can be characterised as research *in* law where the body of law already in existence is interrogated for new directions and themes. In this tradition, law is seen as a self-contained body of knowledge and data that can provide its own answers and can develop itself through the published judgments of tribunals. Here, the role of the legal scholar is to search for clues and to make connections between judicial pronouncements in a quest for new law. Blackstone, in his Commentaries, writes that judges are the "living oracles"[10] of the law. Judges do not decide what the law is, nor do they exercise any personal judgments in determining the proper principle to apply to a case. The judges are simply the mouthpiece of the law.

The importance of this approach is encountered in our discussion around smart contracts, in particular, what they are and whether they are enforceable. The black letter law answer to these questions is that they are what the law allows them to be and the same goes for their enforceability. If no stipulations are statutory pronouncements to the contrary, then the judges, or "living oracles" are free to interpret them as they wish in accordance with the principles of contract law.

The alternative approach, socio-legal, is much less constrained by what has gone before and is able to scan the horizons in a more inter-disciplinary endeavour. Socio-legal research has been described as a *"signal for careful, concise, clear thinking about law in all its manifestations."*[11] This advances the view that law is not simply about the internal enquiry about its meanings as set forth in judgments and statues. The meaning of law is investigated in an external enquiry into the law as a social entity.[12] In this tradition, it is only through investigating stakeholder perceptions into any phenomena that the truth is observable. This is about how law is received amongst its stakeholders and their views and perceptions. The value in observing these views is even more heightened when the subject matter is fast paced and polarising, such as the topic of technology. How should law seek to facilitate society's adoption of technology and to regulate our conduct and transactions with one another? It is only by engaging in topical debates that the law can seek to maintain its own relevance and currency. This is not to give the impression that the law is in someway of secondary importance in this connection. Far from it. Law remains, in the author's view, the most important force in society.

The blending of the black letter and socio-legal approaches in this book is not a novel approach. The legal philosopher, Durkheim, maintained the validity in this approach by advocating that a combination of state rules combined with an individual's higher sense of morals and ethics imposed by society itself bred the most effective balance.[13]

Digital innovations and data raise many interesting legal issues and this work seeks to address a selection of these. The objective is to provide a commentary on developments drawn from a range of sources and debate alongside a consideration of how this fits with the legal frameworks both existing and anticipated.

1.2 The productivity challenge for the construction sector

The point of automation is to make life easier for us humans and to enable us to work more efficiently. The potential for machines to provide this service has long been recognised, for instance in this 19th-century quote from John Stuart Mill:

> Only when, in addition to just institutions, the increase of mankind shall be under the deliberate guidance of judicious foresight, can the conquests made from the powers of nature by the intellect and energy of scientific discoveries

become the common property of the species and the means of improving and elevating the universal lot.[14]

The rationale behind this statement is that "scientific discoveries" (for which we can read technological advancements) are of limited use on their own and must be aligned with good laws and clear thought as to how they can be best exploited. The search for the right balance in these three factors is a key theme in this work.

The above quote is a reminder that the benefits on offer pertain to the whole economy and society in which the construction industry is an important component. It is important that construction improves upon its productivity record. Off-site construction and modern methods of construction are gaining traction as the main routes to better productivity. Both of these initiatives present challenges to the way the construction industry operates in terms of its building contracts and security arrangements. Consider, for example, the different payment regime required for off-site manufacture such as advance instalments rather than monthly valuations in arrears.

The construction industry is crucial to the long-term success of the UK economy as it includes more than 280,000 businesses and accounts for 10 percent of employment. Worldwide the construction industry is worth about $10 trillion every year. The industry's reputation is poor in terms of technological uptake with the McKinsey Global Institute ranking it as the lowest sector with stagnating productivity and being the least digitised. Construction's position lagged behind other sectors such as oil and gas, hospitality, mining, education and retail.[15] If the productivity of the sector were to catch up with the total economy then a rise of $1.6 trillion a year would be unlocked.

The global construction industry has grown by only 1 percent per year over the past few decades. Compare this with a growth rate of 3.6 percent in manufacturing, and 2.8 percent for the whole world economy. Productivity has grown 1500 percent in retail, manufacturing and agriculture since 1945. One of the reasons for this is that construction is slow to adopt new technologies. There are signs that this is now changing and it is down to the advisors and academics to help accelerate and consolidate these changes. Law has a crucial role to play in building confidence in new arrangements and persuading the stakeholders that the new approaches are reliable, profitable and predictable. In construction, more than any other sector the writer has observed, the old maxim applies, as people refuse to take up opportunities for advancement because they would rather *"always keep a-hold of nurse for fear of finding something worse."*[16]

The technological state of the construction industry is not sufficiently digitised to take advantage of the next wave of initiatives that will have far-reaching implications. Investment in construction technology has, accordingly to McKinsey, doubled to $18 billion in recent years. Despite this, the prevailing opinion is that the industry is not ready for the level of collaboration and information exchange required to make a success of the digital approach.[17]

The frustration of those trying to promote the need for and pace of change is not hard to find. Mark Farmer, author of the report "Modernise or Die"[18] survey, observes that construction is not one of the sectors that has gone through any form of a productivity or innovation phase and still represents low-hanging fruit for the application of technology and process improvements.

The Government has added its voice to the call for improvements through the Construction 2025 Government strategy containing the long-term vision for a smart, efficient and technologically advanced industry. The Government wishes to act as an intelligent client in order to increase economic growth. The Government partnership with the Innovate UK funding body places data at the heart of the strategy with a programme to transform how the UK construction industry plans, builds, maintains and uses the infrastructure, as well as the renewal, replacement and creation of new built assets. New technology increases the detail, currency and quality of data available for everyone. This in turn offers new insights and answers to important questions and challenges facing the built environment.

There are signs that the construction industry is moving with the times and that there is take up of the digital infrastructure around data availability. In the recent NBS Construction Technology Report and Survey 63 percent of respondents regularly use cloud computing for their file storage and one-third use Augmented Reality, Virtual Reality or mixed reality technologies. Drones are becoming commonplace. However, most of the findings point to what the respondents think will happen in the next few years, less so with the here and now. Mark Farmer considers that digital automation and augmentation of the industry's current proliferation of labour-intensive, low-efficiency processes are now in the cross-hairs of tech start-ups, investors and entrepreneurs who rightly smell an opportunity.

There are multiple reasons for the slow pace of progress. One is to do with the inability to store and share data generated by development projects and to make it available for public use. Efforts to achieve this continue in the United Kingdom with the Cabinet Office data initiative[19] and on the international scene by the United Nations Integrated Geospatial Information Framework. Concerns over ownership rights and cyber security persist and have limited the uptake thus far. The central tenet of this book, is that law has to normalise and sanitise this area of endeavour in order for people to have the required faith in the undertakings. The main legal development required is the smart contract, which is promoted here as the key to the issue of lag and reluctance in the construction sector.

Another driver for automation in the UK construction industry is to counter the shrinking workforce and the challenge to build for the increasing number of young people seeking to establish themselves on the housing market. In the construction industry, the total of workers over 60 has increased more than any other age group.[20] When employees retire, the industry loses their essential skills. Extending the retirement age is not a viable solution given that many construction jobs are very physically demanding. In these circumstances, one does not need to look very far for a gap in the market for automation to assist and to replace manual workers.

Neither is the clamour for construction likely to go away any time soon. The United Nations predict that an additional 2.5 billion more people will live in cities by 2050. This will require a vast productivity improvement with the onus on harnessing automation and on building as sustainably as possible.

The other great issues of the current age – climate change, overpopulation and environmental concerns also stand to be addressed, in part, by technology. Smarter use of scarcer resources and the facilitation of the circular economy can be achieved by data management. The circular economy is a movement towards greater accountability in the sourcing of materials making up the supply chain of a built environment asset. The greater the reliability and traceability of the provenance of a component, the more we know about its carbon credentials, its usefulness as a resource, its replacement span and its disposal/recycling. The "circle" is the connection between a component's end-of-life use, linked to its origins.

"Building Back Better" is another concept where technology has a role to play in ensuring that reconstruction in climate change–effected disaster areas does not waste the opportunity to start afresh in redevelopment. However tragic the circumstances frequently are, we should embed better practice in terms of prefabrication and meeting the needs of the effected communities allowing for regrowth and smarter built environment assets.

Another big idea in this context is the aspiration that those countries in the developing world can use such technologies to leap frog from development, which is harmful to the environment in many ways to more sustainable approaches. The power requirement of these technologies should also not be overlooked. Predictive analysis, data mining and machine learning all draw power that has a carbon emission footprint even when taken from renewable sources.

The ubiquity and global nature of data has the potential to design away borders and lead to lasting benefits for the good of present and future generations. The key here is to make the data freely available to allow the good practice to spread without having pay walls in the way. Modern entrepreneurs are comfortable with this free to use approach (for example, the Accord Project making its smart contract templates open source). However, resistance remains in form of the traditional view that assets must be owned and controlled rather than shared.

Or, maybe not. We should also entertain the idea that not that much is actually happening in terms of change in the real world. Granted, processing power continues to become more powerful and mobile phones use eye wateringly high packages of data. However, the breakthroughs in recent years are not, from this viewpoint, anything like what happened in the previous 50 years. To continue the argument, has space travel really come on since landing Neil Armstrong and Buzz Aldrin on the Moon in 1969? The answer is not really that much. We should not therefore take it as read that this new age is as all pervading as claimed in many quarters. Statements like *The Future has already arrived it is just not yet evenly distributed*[21] are certainly arresting but the truth of the statement is debatable. Critical assessment of the claims being made will therefore accompany the narrative in this book. This is in the interests of balancing the more fanciful notions being explored and ground the proceedings in reality.

In light of this, it is worth stating that it is entirely possible that the current build-up and exhortation for better practice will not result in the changes envisaged in this book. This would be a waste but it would not be without precedent. The arguments around this point are explored through the chapters of this work. The capabilities appear to be present to be able to do so much more in the built environment and yet on the ground the resistance to change remains.

1.3 Changing professional roles

One of the most discernible characteristics of the built environment is the number and variety of professionals involved. From inception to completion of a project, there can be dozens of specialists from planners to ecologists to architects and building surveyors, quantity surveyors, construction project managers and real estate surveyors. The writer has come into contact with all of these through the Faculty in which he works. The Faculty has long promoted its inter-disciplinary ethos in an attempt to break down the silos of professionalism that exist. This section reflects on the nature of professionalism and what this means in the context of the built environment.

The language of the built environment is, for the most part, familiar to all the professionals who operate in it. There are a common set of measurements and values (most of which are now metric), to which most can relate. For example, the price per square metre (£pm^2) and rental income per square metre are part of this shared parlance. However, there are limits to the ability and intent of the professions to work and relate to each other. Each profession has a specialist language, fluency in which is part of membership of the exclusive club. For example (and something to which the author can identify), lawyers love to throw in the occasional expression in Latin. This dates back to the time when the Roman Empire slowly crumbled and disappeared. The new orders in the former empire were left to gradually adapt the Roman Law for their burgeoning legal systems. These are the roots of the Common Law in England and Wales and the Latin maxims and phrases are a legacy from these times. The author also has a lingering respect, dating back to an "O" level in Latin,[22] for this ancient language and its inherent beauty and mysticism. The phrase, which might sum up the theme of this book, is *Deus Ex Machina* which is more evocative than its english translation, a god from the machine.

Elsewhere in professional parlance acronyms and initialisations are intended, at one level, to exclude and obfuscate. Clients and fellow professions in related specialisms can find the terminology infuriating, elusive and exclusive. One of the standard fault lines in construction has often been the relationship between the architect and the mechanical and electrical engineers. Both professionals are usually wary of being exposed to liabilities for the other's actions and mutual antipathy can sometimes ensure. The hope is that these barriers to working together productively are likely to be lowered in the coming years as data sharing encourages standardisation and process analysis.

Whichever profession and its language is being considered there is a common challenge – how to respond and/or stay ahead of technological enhancement. Depending on one's age and aptitude, this will be seen as either a threat or an opportunity. One work which considers the changing nature of professionalism in some depth is the *Future of the Professions* by Richard and Daniel Susskind to which this section owes some of its references. These authors continue the traditions of a long line of writers tackling the vexing question of automation. John Maynard Keynes, a noted authority on the subject who gave his name to an approach to economics, answered the question pessimistically as long ago as 1930 "*What will be left for humans to do in the way of work? Less and less.*"[23]

The machines have already made great humanoid-like strides into the physical world by replacing roles previously the preserve of manual skill and dexterity. Here, we encounter the term "Robot" coming from the Czech word "robota" meaning drudgery or servitude. In-built in this definition is the sense that this is the work that humans would rather not do themselves. In the construction context, there is the added impetus of people not being around to perform the work. The skills shortage in the United Kingdom is particularly acute with four people leaving the sector for every one that joins. The need for robots and off-site manufacture are pronounced in this context. Robots, then, perform autonomous roles typically going about their business without human intervention. The early applications of this growing trend is represented by site surveying robots and brick-laying mobile machines. The improvement in health and safety records is an important benefit here.

People tend to be extremely unforgiving when chastising robots for any mistakes they may make. A tolerance error in a robot-constructed house might have the professional team rolling their eyes whilst overlooking at least a comparable rate of snagging issues on a comparative piece of work performed by humans. The saving grace for the computer system though is that it has the capacity for machine learning – it can learn from its mistakes and does not repeat them.

Susskind describes automation as "the comfort zone of technological change for most professionals."[24] This process serves to make the human's role that much easier by replacing the servitude and drudgery aspect of the role with computer assistance. In the built environment context, this looks like surveyors using drones to perform fly-over assessments of railway lines or building exteriors. The 60-year-old brick layer is supported by a hod-carrying third robotic arm, which allows them to overcome any physical limitations. The comfort for the professional is that they consider themselves still "in charge" and rendered ever more capable in the way they traditional services are delivered.

However, automation can also be more transformative. The focus of BIM is to reduce the gap between what the design team have in mind and what the constructors actually deliver. Renzo Piano may be able to convey this in a three-second sketch of the Shard building on a napkin but most constructors and clients would need more to go on.

This current trend for innovation beyond the semi-automation of existing roles represents a more fundamental challenge to long-established practices. The case for their adoption though is particularly compelling when the lower cost, higher quality and quicker time involved are considered. In short, the convenience factor is that much more apparent in technology.

One of the main contentions of Susskinds' work is that *"The keys to the kingdom are changing, or, if not changing, they are at least being shared with others."* The "kingdom" being referred to is the field of professional work enjoyed by any one specialism. The "others" are either competing professionals or newly enfranchised workers able to access hitherto restricted areas of work. More alarmingly for some, "others" also refers to the increasingly capable machines able to assist, augment or replace humans. The 19th century saw mechanisation and labouring jobs being superseded by machines. The "satanic mills" of the North of England represented the impact of the industrial revolution on tasks hitherto performed by hand. The 21st century is predicted to see the encroachment of the machines on tasks that have been the historic preserve of the professions. The newness of these challenges stem from having moved from "blue collar jobs" to "white collar jobs."[25]

From the consumer point of view, this is good news as expertise becomes more accessible and affordable than ever before. It is less good news from the point of view of the professional. Computers are capable of discerning patterns, identify trends and making accurate predictions based on big data. The advancement towards self-executing platforms is also credible. This raises the prospect in the not too distant future that for most professional work a computer system will be able to give a reasoned response or advice, fully documented, to an expert standard.

Competition between some professionals as to who does what best is essentially beneficial for the client as the user of the services. The profession of project managers in construction has sought to challenge the monopoly architects used to have on the contract administration role in construction. Standard forms such as the New Engineering Contract have supported this movement in making the project manager the key professional appointment. Another aspect to the argument is that sometimes professions no longer wish to "own" a discipline and effectively cede the territory as, it can be argued, is the case with architects who appear reluctant to perform contract administration. The gap in the market for automation therefore grows larger – an ageing workforce and some professions willing to concentrate on their core disciplines.

The open-ended nature of the answer to the question "where will it all end with these machines?" makes a good deal of people feel nervous around the so-called rise of the robots. At one level, it is the human condition to not merely accept a state of affairs as fixed. It can rightly be said that what has made humans evolve from other upright apes is the capacity to worry about things and, hence, develop thinking strategies to protect against the non-desired outcome. That is not to say that automation is an undesired outcome, the writer takes a phlegmatic approach to this issue – the genie is out of the bottle and the stopper cannot be re-inserted. People have a choice including taking oneself off-grid or back to basics would be an extreme response to the rise of automation. A more measured response would be to take the occasional holiday from technology and this phenomenon may become a more common event.

The more positive response then, is to seek to "*race with the machines rather than against them.*"[26] This involves accepting the fact that automated systems are better at some bodies of work that are currently undertaken by people. Within the built environment, professionals need to come to terms with deference to the superior capabilities of machines. This can already be seen in construction and property. The key skills most prized by quantity surveyors and real estate surveyors are measurement and valuation, respectively. It is not difficult to see that the basic operation of these tasks can be partially, if not wholly, automated. In the first case, running a slide rule over printed plans has become an anachronism replaced with measurement software compatible with BIM. In terms of valuing property, relying solely on an agent's gut feeling is now effectively bypassed by online valuation databases.

The challenge for the professions is to recognise the opportunities for better technologically enhanced service and offer these roles to their waiting client base. An example of this is to recognise that "take-offs"[27] are machine generated but that the funder of a project still desires the assurance of the professional that these have been performed accurately and correctly. Offering this enhanced auditing service, backed by the appropriate level of professional indemnity insurance, ensures continuing market share and professional work. In the final analysis, access to legal recourse for negligence actions against their professional team and their machines remains the bottom line for the client and their funders.

Reference was made earlier to retaining a degree of scepticism towards the claims of how fast things are evolving and how great the risk of being left even further and further behind. This, in part, stems from the media where we hear of the exponential rise in the processing power and capacity of the underlying technologies. However, it would be wrong for any profession to seek to draw comfort from having struck a balance with technology whereby it will encroach a certain distance but no further. The implication is that the hinterland, to which a profession may have retreated, safe in the knowledge that this part of their role is inalienable, is in itself also now automatable. A phrase used here is between "routine" and "non-routine" work and the extent to which the former is routinisable. This may lead nervous professionals to enquire, "is nothing sacred or safe from the reach of the machines?"

Further, a machine may take a different approach to the provision of a service that renders the human approach un-necessary. This could involve a statistical approach rather than a reasoned approach to such things as valuation. The valuer's "intuition" on what property is worth becomes less valuable when an algorithm can value based on 1000s of similar transactions.

Returning to the Susskind work, the main stay of their book is that there are three forces at work that professionals need to endure and, if possible, embrace in the technologically enhanced future. They are:

1 Disintermediation – this can be defined as the removal of the intermediary or agent or broker from transactions. For example, one of the massive benefits of a blockchain approach is the removal of the trusted intermediary such as a bank or record keeping institution such as a registry. The very notion of handing our money or data to a third party for safekeeping will, in time, itself appear odd to those brought up on peer-to-peer and decentralised ledgers. In the built environment context-cited examples include architects being disintermediated by Computer Assisted Design and BIM. The Quantity Surveyor is similarly displaced by technologies such as Bluebeam and CostX.

2 Reintermediation – the disintermediated professional faces a stark choice between survival and extinction. One route to survival is to find another home amongst the professional habitats available. This does not necessarily involve doing some other profession's role but potentially laying claim to a new role potentially made by the technological advances themselves. Ten years ago, there was no such professional role as a BIM manager but now many students are signing on to a conversion master's programme to avail themselves of these new professional opportunities. Another example of new markets being created by technology involves drone delivery in the City of London. Roof spaces on skyscrapers has, previously, not had much in the way of material value. However, as drone package delivery becomes a reality, the real estate market in roof space and the airspace directly over it develops its own status as a trading commodity requiring professional service from Roof Top Real Estate surveyors. A third prediction is the digital asset manager. Clients will need to establish how they are going to maintain their digital data, how frequently and who will take on the responsibility. The Hackitt report identified the need after the Grenfell Tower disaster to ensure an information trail and decision log throughout a built asset's life.[28] For digitisation to have long-term benefits, professionals will need to focus on the legacy value of data.

3 Disaggregation – this process is likely to be taking place regardless of whether one's role is becoming disintermediated and whether or not one is seeking a reintermediation. Disaggregation refers to the breaking down of a lump of professional activity into its constituent tasks. Those tasks are then assigned to those best placed to perform them at a cost that will be acceptable to the client. An example of this can be seen in the field of legal services. Road Traffic Accident claims, previously performed only by solicitors, become broken down into tasks performed by paralegals. This is possible because liability is frequently incontestable by the party having caused the accident and the claim simply involves a reckoning of quantum or the amount of damages. The solicitors' role is disaggregated to the extent that it has partially been stripped away in relation to quantum claims whilst they remain involved in

liability claims. No great foresight is required to see that the paralegal role may be disaggregated to the realm of automation except for the saving grace that people like to deal with people – or at least machines that sound like people.[29]

One of the knock-on effects of disaggregation will be an increased focus on the outcomes of professional work rather than the time spent in delivering it. The lawyers "hourly charge" has frequently attracted attention as an unpopular means of billing a client. The counter argument was always "how long is a piece of string?" in that the lawyer could not supply an exact quote for a piece of litigation with large variables to contend with – such as the approach of the other party or the work involved with complying with the directions the judge dictates for the proceedings. The benefit of the technological future envisaged here is that the degree of uncertainty of outcome and approach will be much reduced as things become standardised.

Many professionals may fear that this sort out standardisation leads to the imposition of a "one size fits all" approach. In fact, the opposite may be true in that the greater level of granularisation allows for a more individualistic approach tailored to the exact needs of the situation. Actual customisation is possible. For example, 3D printing on a building site can supply there and then the exact component needed in a building services component. This can be a very valuable solution to a supply need, particularly in remote locations – a 3D printer has been sent to the international space station. Looking further ahead, the terraforming of Mars, from the red planet to a green planet is likely to require some substantial 3D printers.

To conclude this section, it is worth remembering that Susskinds' disaggregation does not seek to remove the human from the equation altogether. The benefits are to give the recipient of the work insight into what was involved in delivering the role. This greater understanding of what is involved, and the training and experience required to give advice and recommendations, could result in a better and clearer appreciation of a professional's contribution. To whatever extent humans are technologically enabled, the paying party is likely to remain time poor and have a large number of competing claims on their limited attention. Paying a professional for a service will remain the go to option for the many. However, for the few who recognise that it is the data not the service which is important, the words of Tim Berners-Lee, founder of the world wide web, will resonate: *"Data is a precious thing and will last longer than the systems themselves."*[30]

1.4 The importance of data

In recent years the generation, capture and analysis of data has grown rapidly in other sectors of the economy. This has allowed for machine learning–based approaches aimed at analysing how tasks are performed and how they can be optimised. However, building projects are routinely still seen as a series of one-off projects and the learning available from a study of completed projects is squandered. Partnering, or alliancing has sought to keep the learning and relationship building from within a project together for the benefit of future projects. Preserving the team ethos is still a relatively rare occurrence in construction even inside a framework of alliancing contract. People move on and are appointed on different projects resulting in a situation where the project know-how and interpersonal relationships carefully nurtured between the team members are lost. However, following on from the Berners-Lee quote, it will ultimately be the value of data and not the culture of the project which will have long-term importance. Conversations around the construction industry reveal the sense that data gushing out of the ground like the new oil and the sector has not invented the combustion engine. The comparison to oil is not wholly accurate. Oil pollutes and is a finite resource. Data, on the other hand, appears to be a limitless commodity and does not pollute. However, it should always be borne in mind that processing power requires electricity that does have a carbon footprint. The huge electricity demand of industrial scale "data mining" is a reminder that data should be used proportionately.

The detractors of the collaborative approach have the contrary view that the longer a construction team stays together the more the feeling may start to grow on the client side that the supplier is becoming too comfortable. Steve Morgan, the former Chief Executive of British Airway's Authority, once memorably said that he had to win his business from his customers every day – why should he treat his suppliers differently? This has been the tension at the heart of the partnering agenda reviewed in section 7.2, where the limits of the common purpose are exposed.

The most valuable commodity squandered when a team breaks up is the buildability and knowledge of each other's' work and working practices. The most essential commodity here is the trust. "Trust?" One main contractor once said to a client's representative in a seminar, "I trust you to rip me off!" Whether complete trust in a commercial setting is ultimately achievable, remains a moot point.[31] However, collaborative behaviour can be embedded, if not subsumed, into a technologically enhanced data driven approach. The blockchain and smart contract systems covered in this book are sometimes described as "trustless" by which it is meant that one party is not required to trust another with whom they are conducting a transaction. A trustless technology is so secure and smooth in handling the business that there is an extremely low risk to sharp practices. The paradox for the collaborative agenda is that trust is best created where it is rendered un-necessary.

The rise of the machines and importance of data is nothing new. What is novel is the capacity of the machines to deliver on the promise. The standard approach to technology has been that an industry would articulate a need and then the solution would be developed. This is no longer the landscape in which construction operates. The technology is waiting for the industry to catch up. And yet, uptake will remain slow. Collecting and analysing large volumes of data requires huge investment in capable systems. This investment simply does not happen in a sector with a poor record for research and development. The market leaders in technology such as Trimble and Revett can demonstrate their wares and promote isolated best practice whilst the rump of the sector effectively stagnates. These companies recognise the value of their data sets and their asset libraries. Those that operate at the vanguard of the construction industry are also alive to the importance of data.

1.5 Contracts as the key

The contract has long been the medium through which business is delivered. The question has already been posited about whether the contract is being used far beyond its limitations when applying itself to the digital age. There appears to be an enduring quality with its simple rules of formation/execution and certainty. The writers of smart contracts appear to realise this and have themselves been smart enough to frame their procedure in analogous processes. For example, the ethereum blockchain token issue involves offer/acceptance and consideration protocols.

Contracts therefore remain essential to give business and people the comfort they need in their dealings with each other. The contention therefore is that a shift is required in how we think about contracts so that they can better apply themselves for technological advancement. The good news is that the very simplicity of contract law means it can be re-imagined without too many divergences from the contractual landscape and existing more innovative forms of procurement. Both require the protection of law over their transactions and this is where the smart contract may prevail. Smart contracts have the potential to bring the legal benefits of clarity and remedy for breach of contract that are fundamental to economic function.

The benefits of the smart contract (Figure 1.2) appear in this graphic adapted from docusign.[32]

The two axes of the graph are a simplified life and the acceleration of business. These have been the mission statement of Docusign's e-signature technology since its inception in 2003. In this analysis, the "wet signed" paper contract is portrayed as a drain on simplicity and an impediment on business. Seemingly endless rounds of negotiation are entered into even before the logistics of actually signing a contract are tackled. Amendments are commonplace as is the stress caused to all concerned in having an incomplete picture of information about the project. Delays are caused by missing elements of the design, commonly the design of sub-consultants and sub-contractors. These issues frequently lead to the use of letters of intent and other potentially dispute engendering temporary measures. The hope is that projects can become more expedient and efficient by employing digital contracts.

The benefits of smart contracts start with the removal of the need for execution of wet contracts. However, the delays in being able to sign a fully formed contract are usually more to do with outstanding information or permissions. The next level of innovation is therefore to connect contracts and projects together, typically by way of a framework arrangement, featuring pre-agreed terms replicated

Figure 1.2 Smart contract benefits

over numerous projects. The Framework Alliance Contract FAC-1 is an example of this, which is reviewed in section 9.3.

A repeatable framework approach has the potential to remove the negotiation and incomplete information aspects of traditional practice. The availability of digital and connected contracts removes these obstacles and makes it much easier to proceed to the signed version of the contract. In the connected environment, this can extend to all members of the supply community.

The ultimate benefit of smart contracts as demonstrated in the graph is to introduce some aspect of automation to the arrangement, either in the formation or the performance of the contract, or both. This was performed in a recent collaboration between DocuSign and Clause[33] where the contract becomes smart and was able to self-execute on the resolution of any issue preventing an earlier sign-off of the contract. This enablement of "smart clauses" means that a contract can respond and, for example, release payment in a timely manner. The smart contract approach would produce an audit trail and immediate transparency for all the parties with access to the contract.

The limitations of the existing contractual arrangements and the case for smart contracts are considered in chapter six. The organisation of the book is set out in section 1.7.

1.6 The sections explained

1.6.1 *The as-is position*

This section chronicles the current limitations of the construction industry (chapter two) and its legal arrangements (chapter three). The approach does not seek to labour the point or attribute any particular culpability for the state of affairs. However, on the basis that a problem well defined is half solved, it is important to set out the position, as it is generally perceived by industry commentators.

The limitation section starts with what it termed the "goldfish memory" problem before examining the contributions made to the situation by under capitalisation. The section continues with a discussion around disputomania by which it is meant the prevalence of disputes within the sector and the industry it supports. The "flashes in the pan" is a reference to short-term thinking within the industry and the patch track record of implementing change follows on from this. Chapter two concludes with a longer look at the contention that professions are ceding territory in the face of the complexity and legal liabilities that abound in the current climate.

Chapter three starts with a reminder of some of the basic tenets of the legal approach including the focus on the negative instruction – for example, do not be negligent and do not breach contracts. This is followed by a section on contract drafting and its inbuilt imperfections. The chapter concludes with a discussion around gaps in the legal infrastructure. Underpinning this is the notion that construction law is not set up for digital age and is being used beyond its limitations. Covering these developments serves two purposes as an appreciation of the "state of the art" legal solutions set up to cope and to build a platform for the study of the future developments introduced.

1.6.2 *The to-be position*

The "to be" position takes the best practice of today and examines the further developments which are expected. In many cases, the technology exists but the infrastructure to support it is under developed. Chapter four introduces the smart contract, at the heart of this book's recommendations, before examining the phenomenon from different angles. Chapter five embeds the predicted developments in their socio-legal context by examining the thoughts and perceptions of industry stakeholders. Chapter six completes the section by looking at the legal basis for smart contracts and related issues.

1.6.3 The steps in-between

The advances made in best practice and finding their way into popular usage appear in chapter seven. This section pulls back from the outer edge of expected developments and looks at the more pragmatic question of what the steps are that are most likely to make progress towards the future(s) outlined. On the assumption that the future laid out is a desirable one, and the momentum appears unstoppable, this section plots the course for how to start the journey. Chapter seven examines the best of the current crop of legal initiatives potentially providing stepping-stones. Chapter eight examines the background provisions needed to support the development of smart contracts. Chapter nine specifically considers the pathways towards smart contracts and the involvement in BIM in this journey is covered in chapter ten.

1.6.4 Online dispute resolution and smart contracts

Dispute avoidance and resolutions are separated from the bulk of the work and treated as a separate section here. The current dispute resolution options are discussed before the contribution that technology can bring to area are considered. The approach of contrasting the AS IS position with the TO BE position and identifying some helpful STEPS in-between is repeated in the context of this question. The section concludes by asking a simple question – if we identify and then remove the triggers of dispute within the construction industry then will we have succeeded in removing the disputes?

1.6.5 Conclusions and the pace of change

A short final section pulls together the threads of the work to a concluding point. Speculation and extrapolation around the pace of change and likely progress are made based on the book's contents. Observations follow such as the new departures in technology and law should recognise the contribution of the existing structures and ensure compatibility and complementarity.

Notes

1. Radin, M. (2017) The Deformation of Contract in the Information Society. *Oxford Journal of Legal Studies*, 37 (3): 505–533.
2. Joel v Morison (1834) EWHC KB J39 A frolic is a word for moving about in a cheerful and lively way, for example, new born spring lambs.

3. Hadley v Baxendale (1854) EWHC Exch J70.
4. Carlill v Carbolic Smoke Ball Company (1893) 1 QB 256.
5. https://www.mckinsey.com/industries/technology-media-and-telecommunications/our-insights/what-digital-really-means.
6. Saxon, R. (2013) *Growth through BIM,* Construction Industry Council, London http://cic.org.uk/publications.
7. Alfred North Whitehead as cited in the *Growth through BIM report.*
8. Chess has had its own Artificial Epiphany two in 1997 when World Champion, Garry Kasparov, was beaten by IBM's computer Deep Blue.
9. Turing, A. (1950) Computing Machinery and Intelligence, *Mind,* 59 (236): 433–460.
10. Blackstone, W. (1829 [1765]) *Commentaries on the Laws of England,* volumes I–IV. London: Sweet & Maxwell. The re-occurrence of the word "oracle" is noteworthy. The word oracle comes from the Latin verb orare "to speak" and were thought to be portals through which the gods spoke directly to people in Greek antiquity. In the world of computer science, an oracle machine is an abstract machine used to study decision problems. This is sometimes called a Turing Machine.
11. Bottomley, A. (1997) *Lessons from the Classroom: Some Aspects of the "Socio" in "Legal" Education,* in Thomas, P., ed Socio-Legal Studies Dartmouth Publishing Company, pp. 163–184.
12. Chynoweth, P. (2008) Legal Research in the Built Environment in: International Conference on Building Education and Research (CIB W89 BEAR), 11th–15th February 2008, Heritage Kandalama, Sri Lanka.
13. Durkheim, E. (1984) *The Division of Labour in Society.* London, Macmillan.
14. Mill, J. S. (1848) *Principles of Political Economy.* London, Parker.
15. McKinsey (2017) *Reinventing Construction: A Route to Higher Productivity,* McKinsey & Company.
16. From the poem "Jim" by Hilaire Beloc.
17. Mcnamara, A. & Sepasgozar, S. (2020) *Developing a Theoretical Framework for Intelligent Contract Acceptance,* Special Issue Construction Innovation: Information, Process, Management, Emerald Publishing.
18. Farmer, M. (2016) *The Farmer Review of the UK Construction Labour Model,* Construction Leadership Council, National Institute of Economic and Social Research.
19. www.data.gov.uk [last accessed 18 May 2020].
20. Chartered Institute of Building (2018) *The Impact of the Ageing Population,* available at www.ciob.org.
21. Gibson, W. (2003) The Economist, December 4.
22. Latin classes at the author's school involved a strict hierarchy of seating arrangements based on your performance in the last test. The well-liked Latin Master, Mr Bennett, also Colonel in Chief of the Cadet Force, no doubt was inspired by classical literature in taking such a meritorious approach.
23. *John Maynard Keynes technological unemployment* (1930).
24. Susskind, R. & Susskind D. (2017) *The Future of the Professions: How Technology Will Transform the Work of Human Experts.* Oxford University Press, Oxford.
25. This is a reference to work attire – white collar refers to professionals (shirt and tie) and blue collar workers who can get their shirts dirty through manual work. The distinction may no longer hold true as professional attire has now adapted and may be more casual than site wear.
26. Bryonjolfsson, E. & McAfee, A. (2014) *The Second Machine Age.* W. W. Norton Publishing.
27. The traditional role of the quantity surveyor is to "take off" the quantities from an architect's drawing and enter the same into a bill of quantities, which can then be priced by the tendering contractors.

28. Independent Review of Building Regulations and Fire Safety, 17th May 2018, Ministry of Housing Communities and Local Government.
29. It becomes increasingly difficult to tell the difference between a human on the other end of the telephone and a chatbot. Scientists have developed an "emotional chat bot" which is fairly convincing as a human.
30. Berners-Lee, T. et al. (2006) *A Framework for Web Science (Foundations and Trends)*. Now publishers, Hanover, USA.
31. See chapter 7.2.
32. Docusign smart clauses available at: https://www.docusign.com/partner/smart-clauses [last accessed 9 July 2020].
33. https://www.artificiallawyer.com/2019/03/14/docusign-can-now-trigger-clause-smart-contracts-co-preps-for-paid-clients/.

Section II
AS IS

2   Limitations of the construction industry's approach

The difficulty lies, not in the new ideas, but in escaping from old ones, which ramify, for those brought up as most of us have been, into every corner of our minds.[1]

The case has been put forward in the introduction that the technologically enhanced future is a desirable outcome. Before setting out what this future might look like, particularly with regards to smart contracts, and prior to plotting the steps necessary to deliver it, a review of the current limitations of the construction industry and its law should reveal the starting point for these developments and allow an assessment of the barriers in the way to adoption. This section looks first at the construction industry's approach, admittedly with a critical eye. The writer is a reluctant critic as it is a truism to say that it is easier to knock something down that it is to build it. Nevertheless, any compromise or acceptance of the status quo potentially leads to stagnation and back to the quotation starting this section. The remainder of the section examines some of the features of construction law which will require a re-purposing to facilitate change.

Collaboration has long been touted as the potential saviour of the construction and civil engineering industries. The sector has been implored for many decades to improve its attitudes and to end its adversarial approach. The disappointing reality is that disputes are as rife as ever. Positive and productive relationships are routinely compromised as a consequence.

A survey a few years ago of final year quantity surveying students at the University of the West of England revealed a baffling outcome to a question. When asked whether they would prefer the construction industry as it is now – red in tooth and claw – or collaborative, they chose the former. I repeated the exercise more recently substituting "collaborative" with "digitally enhanced" and the response was overwhelmingly the latter. Therein lies the nub of the issue. Most professionals are aware that the collaborative approach has huge benefits but nevertheless remain sceptical at least until they have seen it work, and sometimes longer. However, no one wants to be seen to stand in the way of progress. The argument developed in this book is that the emphasis should be changed completely. It is arguably the time to focus on the remorseless progress of technology rather than promoting an ideology that is more social than commercial. The disconnect between what construction industry says about collaboration and the private thoughts of many on the subject sets up an obstacle to progress. There is no such doubts when it comes to technology. This chapter reviews the "usual suspects"[2] of construction industry limitations.

2.1 The goldfish memory problem

Anecdotally, a goldfish only has a memory of three seconds before it forgets everything.[3] The analogy is there to compare this to the construction industry with a similar tendency to forget what it has learnt.

It appears to be generally accepted that the construction industry has a good deal of shortcomings. This can be presented more positively by recognising that there is room for improvement and that there is scope for some quick wins. Neither should we necessarily talk down the achievements of this massively important industry. The built environment plays a vitally important role in our day-to-day lives. Winston Churchill said, *"we shape our buildings and our buildings shape us."*[4] This is observable through the wellbeing of a worker in a pleasant office environment against one in a tired building with limited light and dated furniture and fittings. The former will be much more likely to be imbued with a sense of purpose versus perhaps the more mundane and banal working or living conditions of the latter. The Built Environment also supports our leisure activity. Visits to inspiring architectural feats whether urban or rural, provide us with soul food.

One of the simple needs of humanity is to have some form of shelter. Housing represents a major part of the built environment. Housing policies and preferences change over time. Post-war residential tower blocks were originally greeted with enthusiasm due to excellent views. Deterioration of the buildings made them much less desirable and made them unpopular. This situation was being countered by refurbishment and the popularity for these structures was returning. However, the Grenfell tragedy in June 2017 in the United Kingdom turned attention to the materials specified in high-rise living. Many residents in similar buildings awaiting a refit of their cladding systems and are caught in the invidious situation where they are unable to sell their property pending remedial work. The importance of refurbishment of the existing stock of buildings is a substantial market.

The pivotal importance of the built environment to our lives is therefore self-evident. Its economic profile is of similar importance. All countries celebrate their individualism and diversity by taking pride in their buildings. Quite often though, new "mega" projects attract negative publicity based on their spiralling budgets and time over-runs. Recent examples of this phenomenon in the United Kingdom include Terminal 5 at Heathrow Airport and the Olympic Games of 2012. Terminal 5 at Heathrow was an exemplar of good design and project management that cemented London's status as a global city. The Olympic Delivery Agency in 2012 delivered a suite of stadia leaving East London a legacy of high-quality facilities for sport. At the risk of appearing too London-centric, the skyscrapers (Figure 2.1) themselves cannot fail to impress any visitor with their innovative design and novel approaches to glass and steel.

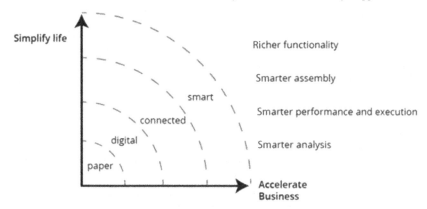

Figure 2.1 The London skyline

Considerable pride and delight exists in looking further back into the history of the built environment and the buildings constructed. However long ago these buildings were built there is likely to have been similar furore about their cost and budget. A poor record in construction management and initial budgeting is a common failing and repeated time and again in a goldfish memory-like scenario. The Palace of Westminster (Houses of Parliament) provides an iconic (if dated) image of the United Kingdom, but featured strikes, delays and a hugely exceeded budget.[5]

The infrastructure needed to host an Olympic games requires a massive investment. Nearly half of Olympic games have cost overrun above 100 percent.[6] Winter games are particularly prone to massive overruns, the games in Lake Placid (1980), Lillehammer (1994) and Sochi (2014) overrun by 324 percent, 277 percent and 289 percent, respectively.[7]

The simple message that ought to be remembered from these projects was therefore to have a clearer grasp on the budget from the start and not allow repeated risings in the budget to give the opportunity of being used to any one party's specific advantage. Recent headlines and spiralling budgets for cross-rail and high-speed networks in the United Kingdom undermine any claims to have made progress on this front.

Nevertheless, on the global scale, it is difficult to overstate the importance of construction and its cousins, civil engineering and infrastructure. Geo-debates routinely centre on such projects as China's belt and road that is expected to cost more than $1 trillion. This represents an unprecedented investment into an industry where the lessons from history indicate that the cost is only likely to rise.

The other major forgetfulness in the industry is in its lack of awareness of its environmental impact. The sinews stiffen at the sheer scale of the concrete and tarmac needed to realise the belt and road project. The wider problem is the predicted rise in population and the consequent pressure of ongoing climate change and environmental damage. Technology and a smarter built environment is key to successful development for future generations. The stakes simply could not be higher in terms of getting this wrong. The emphasis therefore needs to be on gathering the data from the digitally enabled built asset now being created, homes, offices and, in turn, cities can all be operated in a smarter, more efficient and environmentally friendlier way. This data can also be used to assess trends and inform the design of future buildings, infrastructure projects and large-scale city planning.

2.2   Undercapitalisation

Undercapitalisation occurs when a company does not have sufficient capital to conduct normal business operation and pay creditors. This is extremely common in the construction industry throughout the supply community from Tier 1 – large main contractors – to Tier 3 – more modest subcontracting firms. Construction regularly loses more businesses to insolvency than any other sector.

Construction companies have low start-up costs when compared to other sectors and this can exacerbate the problem of a lack of capital assets. The business model of many firms represent a peripatetic approach marking the importance of cash flow, Lord Denning's "lifeblood of industry."[8] Many are run on credit and overdraft facilities, which are often extended to companies on a discretionary basis. One such arrangement memorably left a director of a construction company (an acquaintance of the author) effectively destitute when the bank withdrew the overdraft. "I thought we had a gentlemen's agreement," pleaded the director. "We ain't gentlemen," replied the bank.

January 2018 saw the collapse of Carillion, one of the leading construction companies in the United Kingdom. The corporation has amassed debts of £1.5 billion, which the government refused to bail out. Rudi Klein, remarked that one good thing to come out of the insolvency was the opportunity to peer into its business models. The scant regard for the interests of its suppliers and subcontractors appeared to be the norm. Despite Carillion being a signatory of the Government's prompt payment code meant to ensure payment within 30 days, its standard payments terms were 120 days. A shorter period was available if the subcontractor was willing to take a lesser sum. The House of Commons Select Committee commented, *"this arrangement opened a line of credit for Carillion, which it used systematically to shore up its fragile balance sheet, without a care for the balance sheets of its suppliers."*[9] In short, Carillion was a bank for its supplier's money. A bank with very unfavourable terms for its customers.

2.3   Disputomania

Tony Bingham has been writing accessible and informative weekly columns in Building Magazine about disputes for well over twenty years. In his 2014 Joint Contracts Tribunal (JCT) Povey lecture, he set out his ideas that until the constructing industry discovered its common purpose then it would remain mired in disputes. *"Building is a business where disputes are so very ordinary."*[10] It seems to be generally accepted that disputes are currently part of doing business in the UK

Construction Sector. It has created a very lucrative worldwide dispute resolution market with plenty of work for the claims handlers and lawyers.

Construction contracts by their nature carry a good deal of unknown risk and so contracts are entered into with an incomplete and inadequate understanding of potential outcomes. The loose, flexible and ambiguous nature of construction contracts are perfect breeding grounds in which disputes can proliferate. The manual nature of contract administration contributes towards an inefficient system with poor levels of traceability and auditability due to the onerous manual tasks proving too much on complex projects. Throw into the mix a new or unfamiliar form of building contract and the odds of a misunderstanding leading to a dispute situation are increased. Building professionals tend to stay loyal to the standard form contracts they are familiar with and are often loathe to risk their business on a new-fangled form. The client has to be firm on its choice of contract to force through any departure from the normal routine.

The average amount in contention between two parties in construction disputes involving the clients of just one major consultancy saw a global average of $33 million in 2019.[11] The United Kingdom was $17.9 million. Elsewhere, the average amount was even higher with $56.7 million in the Middle East. The overall causes of the disputes are attributed to failures to properly administer the contract (1) Errors and/or omissions in the contract documents (2) and owner/contractor/subcontractor failing to understand and/or comply with its contractual obligations (3). There is evidently a disconnect between the contracts and their users given that months if not years have been taken to negotiate the terms at massive cost. Another two prominent causes of dispute are defects and translation errors. All of these issues can be eliminated by smart contracts and other initiatives such as off-site manufacture and modern methods of construction.

Apart from the value of the disputes, it is also necessary to factor in the cost of the legal representation. On occasion, the need to recover one's legal fees can itself cause the perpetuation of a dispute. The total expense on what are avoidable situations represents a huge waste in terms of time, money and the human cost of being involved in acrimonious and blame attributing disputes. The possibilities for a better approach to preventing and resolving disputes is discussed in section five. Technology has much to offer in this connection, as it does in all the other areas discussed here in the AS IS reality of the construction industry.

Contracts exist to protect the parties but quite often end up, where they are understood, being weaponised by the users. There is also a high prevalence of spurious claims where one party inflates their claim in the knowledge that they will be bartered down from this position. If contracts are created with a mindset that one party can only do well at the expense of another then the first seeds of disputes have been planted. This mindset is avoided/diminished in collaborative approaches and is arguably not required at all in a technologically enhanced project in a decentralised and highly granular approach to construction.

Even where the parties manage to avoid a conflict situation, the overruns remain alarming common. Construction projects have an 85 percent chance of over-running considerably.[12] The waste involved here is massive and the decisions

Figure 2.2 The mushroom effect as experienced by many a subcontractor[13]

on whether or not to proceed would have been much more balanced if a true picture had emerged about the real cost in time and cost. The challenge then for the industry and its relationship with technology is writ large – bring about a reality where projects are much more likely to come in on time and on budget and thereby reduce the disputes that blight the industry.

2.4 The flashes in the pan

A flash in the pan is a reference from gold prospecting in the United States in the 19th century. The gold prospector would "pan" the streams running out of the country where gold deposits were known to occur by allowing mineral rich water to run through their sieve or pan. Quite often, particles of what appeared to be gold would be noticed only to find the material was pyrite or "fool's gold" which was relatively worthless against what they were actually looking for.

The reference to gold mining is strangely relevant today in the age of crypto-mining. In a further coincidence, some of the Old West towns which saw the original gold rush are being used by crypto-miners seeking cheap electricity to power their computers which perform the mining.[14]

The construction industry has a tendency of treating new developments as if they were fool's gold and disregarding the new discoveries before properly examining their true worth. This is observable with moves towards improving the procurement and management of projects. Most of the technology already exists that could bring meaningful change to the industry and is ready to use. Building information modelling (BIM) is well established in construction and yet its benefits are not universally employed. This is partially down to vague contract provisions and disjointed procurement practices. These extremely capable systems require precision in contractual exchanges and ways of working where data is captured and analysed through the full project lifecycle. Without the holistic approach required, the construction industry grows impatient and seeks to pan other streams.

Attributing blame for this phenomenon to the stakeholders in the construction industry is unfair. At a systemic level, the industry suffers from not pursuing additive, or iterative technological development. One "thing" ought to follow another, for example, smart contracts should follow on seamlessly from distributed ledger technology as they have a common genus. The stop/start faddism of new trends within the industry can be seen to prevent this development. Section four discusses the steps in-between the as is position and the to be position and the principle employed is to identify additive and complimentary advancements.

There is a feeling that BIM is already viewed as "old news" rather than being embraced as a portal to wider digital adoption. The challenge is to embed BIM in working practices. Other industries articulate a need for new technology and embrace it readily. In the autonomous vehicle market, for example, performance is achieved through prototyping and improving on the technology they have in an iterative design and use process. In the UK industry, the temptation is always strong to investigate new fads in preference to the development of innovations that have been proven to work. This conservative approach has at its core the desire to just get on with the job effectively, rather than doing something innovative that might carry some perceived risk. This can be deemed the "just make a start" tendency.

This desire to carry on regardless also misses out the huge potential for integration of the capital project with the facilities management. This often frustrates the main purpose of procurement – to appoint the best value contract over the lifetime of the project.

The construction industry is not alone in taking a short-term view of developments and discounting these as flashes in the pan. A rather more positive long-term spin is put on this by the Gartner hype cycle. The "hype cycle" divides technological advances into those enjoying a technology trigger through to a peak of inflated expectations followed by a trough of disillusionment.[15] Happily, these setbacks are then followed by a slope of enlightenment and a plateau or productivity. The hype cycle has been criticised for a lack of evidence and for having misleading terms, as every technology does not follow the same pattern.

This is to the miss the point that the hype cycle allows us to marvel at the mysterious emerging technologies some of which apparently sink without trace never to be heard of again. The hype cycle also draws on the work relating to techno-economic paradigm shifts.[16] This body of work may seek to establish that even technologies that do not directly succeed still have their role to play in future-shaping.

Smart contracts are represented as being in the first stages of its development where academic enquiry is loudest and only conference organisers are in a position to actually make any money. In the latest iteration of the cycle – 2019 – they had (alarmingly!) disappeared from the graphic as the short attention span, which is inherent with anything involving hype, seeks its next candidate for five minutes of fame. Hopefully, there is sufficient cogency in the arguments included in this book to make a compelling case for the widest possible adoption of smart contracts in construction and provide an example of where improvements in productivity can be experienced.

2.5 A patchy track record of successful government intervention

The UK Government makes regular interventions through initiatives for the construction industry. In little over a decade during the early years of the 21st century, there were 25 different initiatives. Since then, there has been evidence of a more concerted approach as the Government seeks to make good on its promise of being a better client for the industry.

The pattern of government interventions into the construction industry is well established and detailed. The government usually commissions a report from a well-respected industry figure and asks them to write a report into the ills of the industry and suggest some solutions. The reports have unerringly succeeded in identifying the problems, as has been introduced in this chapter, and made extremely sensible recommendations for improvement of the situation. However, diagnosis is not the same as cure. The early reports were allowed to gather dust on government department shelves as the appetite for statutory intervention was evidently not strong enough to administer the medicine required. It was not until the mid to late 1990s that the reports actually resulted in action. The Housing Grants, Construction and Regeneration Act of 1996 was both surgical and dramatic in its intervention. Interventions since have been more measured and used best practice ideas rather than the invasive surgery mandated back then.

The Government Construction Strategy is still current although it dates from 2011. The central notion is that the government should be a "smarter client" and be better able to drive out waste and to unlock innovation and growth by using its purchasing power to drive change. Increasingly, the government seeks to set the best example possible resulting in large framework which are now being operated across government departments in a massive joined up approach to procurement. The desire to try new models and supply chain management contains some notable initiatives including: the requirement to ensure that subcontractors are paid within 30 days, the use of project bank accounts and the two-stage open book procurement approach. The use of BIM is a central theme in the government strategy.

The first major broad-based report into construction in the United Kingdom was published during the Second World War by the Ministry of Works. The Simon Report[17] called for a more collaborative approach to design and construction. The next government report was known as the Emmerson Report dating from 1962.[18] The report identified a lack of cohesion and adversarial, and conflict-ridden relationships were the norm.

These early reports had therefore drawn attention to the fragmentation and lack of consistency in the approach in construction contract law. This was an observation repeated again in the next government report written by Sir Harold Banwell in 1964. The report re-iterated that the most urgent problem with the construction industry was the "necessity of thinking and acting as a whole" with attitudes and procedures needing to change but also suggesting that such changes would be "of no avail until those engaged in the industry themselves think and act together."[19]

The Latham Report recommended legislative change rather than mere exhortation towards better practice. The report raised the curtain on the Housing Grants Construction and Regeneration Act 1996 (as amended).[20] The rationale for the Act was that reluctant and recalcitrant parties had to be ordered to comply with the new regime aimed at transparency, payment notices and a fast-track dispute resolution process known as adjudication.

A few years later in 1998, the later government reports returned to a best practice rather than mandated approach. Sir John Egan's watchwords were about integration and cultural change through partnering in seeking to improve the trust part of the equation. The influential Constructing Excellence movement has its origins in the Egan Report.

The Wolstenholme Report from 2009 was entitled "Never Waste a Good Crisis."[21] This report borrowed a line from President Obama's notion that when things were bad, as they indisputably were for construction during the recession, a good opportunity exists to make long-lasting changes which will be beneficial for those concerned once the recovery is underway.

This report took a reflective look at the journey of the Latham/Egan agenda to date and issued a challenge to the next generation to take up the responsibility of delivering real change. The digital dawn had still not broken and it would be left to the next round of initiatives to seize on this opportunity.

The afore-mentioned Tony Bingham 2014 paper contained an insightful review of the government reports and echoed the findings of the Latham Report,[22] which observed that the two vital concerns in the construction industry were trust and money, and that they were "totally interlocked." The industry has too little trust and not enough money. The panacea suggested was for main contractors to be treated fairly and be able to make a reasonable profit. The parties would do this if they could establish a common purpose.

Figure 2.3 Bingham's common purpose

The optimum situation is to have the two circles overlap as much as possible as this would result in satisfactory trust and money arrangements. The partnering movement adopts this approach and builds on it to produce a model of collaboration. However, there are limitations on how much of a common purpose is possible between commercial entities and how long this can endure for. Businesses exist to be responsible to their shareholders and with a purpose to maximise profits. However, efforts to achieve a greater singularity are extremely worthwhile.

The Farmer Review represented a change of tack by the government bodies. The exasperation of the policy makers reached a crescendo with the review's exhortation to "Modernise or Die."[23] The Review was closely linked to the Government Construction Strategy 2025 that sees collaborative moves based on partnering being replaced with the realisation that innovation and technology can provide the breakthroughs in the trust and money stakes. The movement, Digital Build Britain, also has its origins in this sea-change moment.

The priorities for the government in seeking to improve on the trust and money positions have remained constant throughout. What has now changed is the means by which they can be achieved. Technological enhancement can embed the desired improvements to the extent that trust and fair profits become by-products of the digital approach.

The Hackitt Review Review[24] in 2018 followed the tragedy of the fire at Grenfell tower where 72 people lost their lives. The lowest-cost approach to tendering and procurement approach was identified as a major contributing factor in specifying combustible cladding panels. Dame Hackitt chastised the existing arrangements in the construction industry as a "race to the bottom." Another contributory factor was the non-payment of invoices and consequent cash flow issues causing subcontractors to substitute materials purely on price rather than value for money or suitability for purpose. The Review identified what was missing was the need for a golden thread of good quality information.

The Farmer and Hackitt Review provides important signposts to where the industry needs to go to be fit for purpose in the 21st century. However, both need only look at the fate of most other government reports to know that the work is far from done.

2.6 Not being good with change

Collaboration has been the central message of the government reports referred to in the previous section. By 2005, various industry groups and bodies had come together to create Constructing Excellence, an organisation working across the industry to drive change.[25] The most valuable aspect of the collaboration agenda is claimed to be the ethos of mutual trust and understanding as featured in a suite of new standard form contracts including the New Engineering Contract. Key concepts involved in collaboration include early contractor involvement and the need to communicate effectively.

When reviewing how far these ideas had penetrated into the wider construction industry through collaboration, Sir John Egan awarded the industry a score of four out of ten.[26] Some beacons of good practice shine in the sector but not nearly enough to amount to a critical mass or real agenda for change. There are those that recognise that the "stick" approach is more preferable than the "carrot." Rudi Klein is a commentator who has sought to give the subcontractors in the construction industry a voice. He stated that *"the UK Construction industry never responds to exhortation, official guidance, reports, codes or charters especially when they relate to improving payment security. Such improvement demands a legislative response."*[27]

A similar degree of frustration with the lack of responsiveness is detectable in Sir Micheal Latham's list of those that did not believe in partnering falling into these categories:[28]

- The stick in the mud
- The jobsworth
- The one who just does not get it
- The die-hard sceptic
- The control freak
- The young person fed a poisoned account.

All of these attitudes are presented as being damaging to the industry, leading, as they do to bias, preconception and spurious objections. However, these attitudes are simply manifestations of people's choices in what to believe based on their experience. In an industry with a gold fish memory problem, riven by disputomania, fascinated briefly by flashes in the pan and with a patchy record of government intervention, some attitudes appear entirely justified. The category of young person fed a poisoned account has been something of which the writer has had experience amongst the part time students taught at University. A considerable percentage of the students are on day release from relevant employment. The challenge is always to convince the student to examine their pre-conceptions and to be receptive to new ideas, which they will hopefully promulgate. Ultimately, this comes down to their choice and the constraints under which they work.

Seeking an elusive definition for partnering is one of the challenges faced. Partnering, according to a JCT statement, was more about culture than it was about contracts. At that stage of the definition, a good number of people were already floundering with the intention. The contention here is that digital technologies are much more tangible. The situation arrived at is that collaboration has not worked out as many hoped and technology's impact is similarly barely felt. The proposal in this book is that further progress can be made by putting the two together.

The difference when discussing attitudes to technology is pronounced. Technological change, as we have seen in the section on changing professional roles, does not differentiate between the attitudes of those it comes into contact with. Essentially, if the remorseless progress of technology threatens their livelihood then they will react against its adoption.

Some readers will be familiar with the "oh dear, was that today?" cartoon of two dinosaurs stood on a rock whilst the Ark disappears over the horizon. Despite mixing its imagery (part biblical, part evolutionary), it delivers a powerful message. This is a sentiment that can develop into fear of being left behind.

Management consultants have long employed these insecurities as the hook for their audiences of telling them how far advanced competitor "y" is, compared to them. Similarly, in recessionary times, many a worker has sort, sometimes false, comfort from how the lesser-performing departments are first when it comes to redundancies.

Returning to the list, the response to technology of the stick in the mud, the jobsworth, the one who just does not get it and the die-hard sceptic appear destined to be consigned to history. The control freak and the young person fed a poisoned account should be willing to seize the opportunity given.

The position reached at the end of this review is that there are many reasons to expect further moves towards collaboration through technological enhancement. And yet, constructing excellence and government strategy have been here before. The quote that starts this book reflects that what is needed are good laws and people who respect the law. This then is where to pursue lines of thinking.

Notes

1. Keynes, J.M. (1936) *The General Theory of Employment, Interest and Money*. London, Macmillan.
2. A term used in the classic 1942 film *Casablanca* where the police chief, played by Claude Rains, takes a less than thorough approach to investigating a crime.

3. Recent research has suggested the memory is longer and it may last for up to five months. Either way, in human terms, these are short periods of time.

4. Churchill, W. (1941) Speech in Parliament during a debate about the rebuilding of the House of Commons following a German Air Raid.

5. Construction started in 1840 and lasted for 30 years. The interior decoration work continued intermittently well into the 20th century. https://en.wikipedia.org/wiki/Palace_of_Westminster.

6. Put in building magazine reference.

7. Olympic overruns, available at: https://howmuch.net/articles/olympic-costs last accessed 09/07/2020.

8. *Modern Engineering (Bristol) Ltd v Gilbert-Ash (Northern) Ltd* (1973) 71 LGR 162.

9. 2nd Joint Report of House of Commons Business and Works & Pension Select Committees, HC 769, 16 May 201, para 42.

10. Bingham, T. (2014) *Let's Make Friends with Reality.* JCT Povey Lecture, available at: http://corporate.jctltd.co.uk/wp-content/uploads/2016/03/JCT-Povey-Lecture-2014-for-publication.pdf last accessed 09/07/2020.

11. Arcadis (2019) *Global Construction Disputes Report,* Arcadis Consulting.

12. Flyvbjerg, B., Skamris Holm, M.K. and Buhl, S.L. (2003) How Common and How Large are Cost Overruns in Transport Infrastructure Projects?, *Transport Review,* 23, 1, 71–88, DOI: 10.1080/01441640309904.

13. Mushrooms are kept in the dark and fed on BS.

14. The Digital Human, BBC Radio 4 25 February 2020.

15. https://www.gartner.com/en/research/methodologies/gartner-hype-cycle last accessed on 20/05/2020.

16. Perez, C. (2002) *Technological Revolutions and Financial Capital: The Dynamics of Bubbles and Golden Ages.* Elgar, London.

17. Ministry of Works (1944).

18. Ministry of Works (1962).

19. Banwell, H. (1964) *The Placing and Management of Contracts for Building and Civil Engineering Work.* HMSO, London.

20. Amended by the Local Democracy and Construction Act 2009.

21. Wolstenholme, A. (2009) *Never Waste a Good Crisis,* Constructing Excellence.

22. Latham, M. (1994) *Constructing the Team: Joint Review of Procurement and Contractual. Arrangements in the United Kingdom Construction Industry.* HMSO.

23. Farmer, M. (2016) *The Farmer Review of the UK Construction Labour Model.* Construction Leadership Council, National Institute of Economic and Social Research.

24. Independent Review of Building Regulations and Fire Safety, 17th May 2018, Ministry of Housing Communities and Local Government, para 9.4.

25. Constructing Excellence (2016).

26. Wolstenholme, A. (2009) *Never Waste a Good Crisis.* Constructing Excellence.

27. Klein, R. (2019) *Payment in the Construction Industry. Where are we now?,* Society of Construction Law D220.

28. Mosey, D. (2019) *Collaborative Construction Procurement and Improved Value.* John Wiley & Sons, United States.

3 Limitations in the legal approach

Advances in technology have resulted in a changing landscape for construction contracts. This presents a challenge for legal systems, which are used to acting in a responsive and reactive mode and are based on supplying answers to situations which arise. This time lag between a pressing need and the response can make the law seem out of touch with developments. The lawmakers in England and Wales have sought to remove this limitation and efforts such as the Legal Statement on Cryptoassets and Smart Contracts[1] demonstrate a readiness to encourage legal developments rather than wait in the sidelines. Notwithstanding this, the statement amounts to a signal of willingness and a promise to take an open-minded approach rather than anything particularly tangible.

Hitherto, legal systems have had a good level of success when it comes to promoting commercial activity. The attractiveness of financial centres for business is based substantially on the legal infrastructure it has in place and its ability to provide the security and confidence based on a consistent approach. The English and Welsh jurisdiction has a strong reputation, not only for good laws but also as a well-functioning dispute-resolution centre. The question is whether the law can cope with the technological advancement and deliver a similar amount of confidence to business and personal users as currently exists.

England and Wales has long prided itself on its common law approach. *"From little acorns mighty oaks can grow"* is a popular refrain used to promote the flexibility and adaptability inherent in this system. We need look no further than the infamous case of Donoghue v Stephenson[2] to appreciate that one mouldy mollusc pickled in ginger beer can create a whole new basis of tortuous liability. The common law approach is adopted in many countries around the globe where the doctrine of precedent sits alongside and often complements the statutory body's ability to legislate.

Donoghue v Stevenson (Figure 3.1) represented a major departure for the lawmakers. A customer who became unwell after drinking some contaminated liquid bought for her was able to successfully sue the manufacturer despite the absence of any contractual link between the two and the additional link in the chain in the form of the retailer, Mr Minchella. The over-reaching of the parties in-between the two litigants created the law of negligence. Similar bold "leaps of faith" might again be needed in order for the law to remain abreast of technological developments.

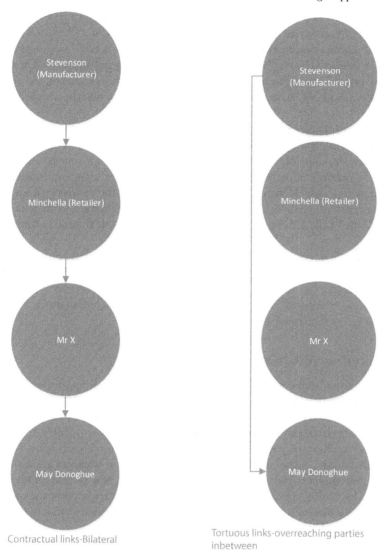

Figure 3.1 The revolution in the law by Donoghue v Stevenson

The mystery persists as to who May Donoghue was with that fateful day, who was Mr X? The legend has grown up that this was a married man who did not want to be identified at any stage in the long proceedings and who was no doubt hoping the case would go away. His hopes could not have been more misplaced!.[3]

The checks and balances built into the (unwritten) constitution ensure that Judge made law is separate from the executive law. This Separation of Power provides protection against abuses. The two sources of law quite often work together to address each other's shortcomings in that a case can plug a gap (or a lacuna) in an Act of Parliament or a new Act can untangle a mess of contradictory cases.

These law-making institutions are likely to require all their cross-support to meet the demand placed on them by the latest technological advances. The wider any gap becomes between the state of the art and the state of the law, the greater the disconnect between the regulative function of the law and the people it seeks to serve.

The twin institutions of the judiciary and executive are not the only offices seeking to bring change. In the sphere of construction law, there is a huge role to play for the standard form contract writers. Their ability to innovate is unbounded by the constraints of parliamentary time or the pertinent facts and issues being in dispute in a case. However, with certain exceptions, the approach of the contract writers has been reactive as it is based on an unwillingness to alienate their users. Many of the latter prefer a tried-and-tested approach than something new fangled.

First, the chapter examines certain features of the law that pre-dispose a reactionary response to developments before examining developments in standard form contracts. The chapter concludes by looking at divergences in national approaches to selected construction law issues and incompatibilities between existing law and technological advancements.

3.1 Focusing on the negative

Law has always been easier to operate when it focuses on telling people what not to do. The guiding principles in contract and tort are essentially the same – do not breach your (contractual or tortious) duties or else you may be pursued for the resulting damages. The purpose of contractual damages is to put the innocent party back in the position it would have been had the breach not occurred. The collaborative movement in construction law has on occasion attempted to change the emphasis and focus instead on telling people how they should behave. The results of this approach have been mixed. The starting point for such an approach is undeniably sound – what creative ways can be used to seek to improve on the observed behaviours in the construction industry?

This negative/positive aspect matters in the crucial distinction between reasonable skill and care (negative) and fitness for purpose (positive). These duties are reviewed in respect of the crucial role of the lead designer, the Architect.

3.1.1 Reasonable skill and care

Historically, the term architect is derived from the Greek term meaning, "head builder." Arkhitekton was used to describe the leading stonemason of the ancient Greek temples of around 500 B.C. – Arkhi meant head, chief or master and Teckton meant worker or builder. The head, chief or master builder would take responsibility from inception and would mastermind all the construction details and be ultimately responsible for the building delivery. Architecture itself, however traces back to ancient Egypt; Imhotep is said to be one of the legendary heroes of the profession; the prototypical "architect" revered for his great wisdom as a scribe, astronomer, magician and healer, revolutionising construction with the use of stone to create pyramids; yet, the Roman builder Marcus Vitruvius Pollio is often cited as the first architect. He undertook the role of chief engineer for Roman rulers such as Emperor Augustus and documented building methods and acceptable styles to be used by governments. His three principles of architecture – *firmitas, utilitas, venustas* (strength, functionality and beauty) are used as models of what architecture should be, even today. Architects were considered learned persons, entrusted with the prerogative of delving into sacred books, sharing the company of the king, and were powerful administrators of the land.

Invariably an architect, or any other professional for that matter, may make mistakes. As building projects have become more multifaceted, errors and omissions by the professional team become prevalent. The construction industry is litigious by nature, seemingly with its fair share of distrust, unethical practices, stereotypes, disputes and claims. The reputation and numbers of lawsuits/litigations have created pessimistic owners, who try everything they can to have their project meet the three primary objectives of time, cost and quality. Anytime one of these objectives are not met, the owner will seek to question the credentials of the professional team. Consequently, the architect is a popular target, should issues arise. The received wisdom is that the architect has deep pockets, thanks to their professional indemnity insurance, and the insurers have a propensity to settle.

The issue of liability on an architect can come down to, unless the contract provides otherwise, whether they have been negligent. Professional negligence involves an examination of the question "has the architect fallen below the standard of a reasonably competent fellow architect?" If the answer is yes then the architect is liable, if no then they are not liable. Professional negligence claims involve a good deal of time and money when they form the subject matter of litigation. Extended court pleadings, discovery of documents and exchange of witness reports, both factual and expert, are the norm. The role of the expert witness is an interesting one – besides the judge, only experts in a courtroom are allowed to venture an opinion (based on their qualifications). Where it is permitted, each party to the litigation will bring their own expert witness who owes a duty to the court to assist the Judge with the questions and issues within their remit. There have been plenty of instances where parties have sought a "hired gun" to support their case rather than an impartial party giving assistance to the Judge. The legal teams will also arm themselves with long lists of case law to seek to convince the Judge that their case is the one which, on the balance of probabilities, should win the day. The loser then has the further burden of paying for most of the winner's costs.

All the formality and rigour applied to a professional negligence trial stem from the fact that the question posed is highly nuanced and depends heavily on the facts of the case. No two cases are the same and therefore there is ample room for legal argument on how the existing law should be interpreted, let alone on what the true construction of the facts of the case are.

3.1.2 Fitness for purpose

At first impression, the fitness for purpose test appears much simpler than the reasonable skill and care test. Disputants can argue at great length and cost around whether or not the architect acted below the standard of a fellow competent architect. However, if the architect was responsible for the function of a building then it was either fit for its intended purpose or it was not. The legal test appears to have moved to a binary footing. It should not go unnoticed that a binary footing is exactly what a computer programmer of smart contracts would prefer.

However, it is rarely that simple in practice. Fitness for purpose appears to involve an absolute obligation to achieve a specified result, a breach of which does not require proof of negligence. Arguments can arise around exactly what the intended purpose was and the extent to which it was met.

Most architects (and design and build contractors) approach fitness for purpose warranties with caution. This is not least because their insurers will not countenance such an approach and refuse insurance cover whilst seeking to dilute the obligation to reasonable skill and care.

The distinction between the two tests came to light in the recent case of MT Højgaard v Eon.[4]

The case involved the failure of the foundation structures of an offshore wind farm in the Solway Firth in Scotland owned by Eon. The structures were designed and installed by MTH. The central issue was which party was responsible for the remedial costs with regards the foundation design in the sum of 26 million Euros.

The initial tender documents insisted on compliance with international design standards J101 which was duly incorporated into the contract between the parties for the design, fabrication and installation of the foundation structures for the turbines. The problem was that there was an error in the minimum standards set by formula in the design standards, which meant that any installation following them would lead to a "dramatic failure." The parties therefore set about remedying the defect before arguing about who should pay.

The distinction between reasonable skill and care and fitness for purpose is worth dwelling on. The former can be described as a lower hurdle for the contractor to surmount – it had displayed reasonable skill and care in following international design standards. However, the higher hurdle, that of fitness for purpose, is more of a challenge for the contractor to establish given the failure of the installation.[5]

The problem in this case was that both tests were incorporated into the contract at different points. There was a general obligation to use reasonable skill and care as well as a technical requirement in the specification that the foundations should last for 20 years. In the first judgment, the specification was held to create a fitness for purpose warranty. MTH appealed and was successful in the Court of Appeal who found that the reference to the specification was "too slender a thread upon which to hang a finding that MTH gave a warranty of 20 years for the life for the foundations."[6]

The decision was again reversed in the Supreme Court[7] where each of the arguments seeking to dispel the fitness for purpose obligation were discounted. All of the judges involved agreed that entering into a contract with dual obligations to exercise reasonable skill and care and a fitness for purpose in respect of a particular result was "not unbusinesslike."

This case represents a step required to usher in the digitally enhanced future. This is the hardening of judicial attitudes around fitness for purpose and away from reasonable skill and care. Where a contractor takes on responsibility for design and construction they become a special class of contracting entity and cannot take advantage of a lesser obligation. Obviously, the smarter solution is not to have contradictory standards in the contract in the first place.

Another helpful clarification to come out of the decision relates to design life and service life. Again, these can be characterised as being a lower hurdle – design life – the design provided can objectively be expected to have a lifetime of [twenty] years, and a higher hurdle – service life – a guarantee that the feature will in fact have a lifetime of [twenty] years. Their Lordships indicated that the design life would be the standard obligation. If the employer required a service life then this needed to be explicit in the contract and the contractor would be able to price the higher obligation accordingly. Moving forward, contractors can take some comfort from the lesser obligation if the fitness for purpose warranty becomes more common place, as indeed it needs to as a step in-between the current and future positions.

The introduction section set out the argument that construction law is stuck in a pattern of stretching existing flawed approaches beyond their limited usefulness. This analogy works equally well in the widening contexts of the built environment, the economy and society itself. Returning to an example of how construction law is warped in places, consider the problem of over-specification that is most common in the Middle East. The nature of decennial liability means that any of the professional team can be sued by any user of the building for any structural fault for 10 years. The solution applied by professionals and their insurers is to massively over specify and build in steel and concrete which performs at many times over the performance than what it will actually require with the accompanying waste of material and environmental harm.

As stated, the smart contract development needs to follow the fitness for purpose as the only logical standard – the thing works or it does not and the party responsible will not be paid. This is not a solution in a vacuum however as merely introducing this test will create ramifications on such areas as insurance and willingness to carry out the work in the first place. Nevertheless, a simplification of this test and provision for limited exceptions. Moving away from reasonable skill and care also makes sense as there is no international equivalency hence why the leading international form FIDIC (Federation Internationale Des Ingenieurs Conseils)[8] contract states fitness for purpose. In the engineering tradition of a contract, this is much more desirable – the power plant has to work perfectly and cannot have a situation things went "a bit wrong" as can be sanctioned by an error making but non-negligent professional.

3.2 Standard forms and contract drafting

The question of whether the current selection of standard forms is currently fit for purpose is addressed in section four of this work. The issue being considered in this section reflects on the purpose of standard form construction contracts and the approach used in their drafting.

Standard forms are primarily intended to avoid the expense and problems of bespoke contracts. To quote Professor Peter Hibberd:

> the whole essence of a standard form is to minimise the transaction costs of enter-ing into a contract, by providing benchmark provisions which aid understanding, by allocating risk in a recognisable way and creating the benefits of precedent.[9]

The same essence is also present in smart contracts. Seeking to deliver on this wish list through the digital medium.

For the standard contracts referred to by Professor Hibberd, the situation arrived at is one out of three has been achieved. Benchmark provisions are pro-vided. Regrettably, this does not aid a detailed understanding and the oft-quoted advice endures to "leave the contract in the drawer until you need it and then hope for the best." Risk may be allocated in a recognisable way but this is com-monly amended for the benefit of the paying party. Precedent is only a benefit inasmuch as it brings clarity. More often than not, it arms lawyers with argument and opportunities for wriggle room within contract interpretation.

A construction risk is defined as any exposure to possible loss or the possible adverse consequences of uncertainty. Every construction project is different and each offers a multitude of varying risks. To ensure the success of a project, a con-tractor prior to starting out on the project must be able to recognise and assess those risks. Sir Michael Latham rightly recognised that risk was omni-present in a construction project and stated, "Risk can be managed, minimised, shared, trans-ferred or accepted. It cannot be ignored."[10] The consequences of poor risk allocation are increased project costs, delays and disputes.

A contract strategy in the private sector usually starts with a checklist of questions based on liabilities and indemnification against the supply side. The involvement of a funder will only serve to heighten the sense that one sided terms and conditions need to be imposed on the providers and the latter are given no opportunity to negotiate away from this position. Often, from the start, adversarial stances are taken and an atmosphere of distrust and unrea-sonable demands further compromises the parties' ability to build towards a common purpose. There are exceptions and contractual negotiations can occur with mutual respect for each other's positions and a well-considered and orderly progression towards executing the contract. Obviously, the luxury of time and having sufficient information are major contributory factors to the temperature of the contract negotiations. Quite often, the negotiating team have neither and the need to force through a contract negotiation in a short timetable only adds to the stresses on the situation. One of the steps in-between discussed in section four is the need for a decent length of pre-commencement time to agree the proper form of the contract or to agree a separate pre-commencement contract.

The process of drafting construction contracts can be very expensive, onerous, inefficient and challenging. Once the client has completed its checklist of the liabilities and indemnification it wants, the next stage is to inform the contractor which of the standard forms it wishes to use. Occasionally, the client may insist on a bespoke contract instead. The most popular standard forms in the United Kingdom are the Joint Contracts Tribunal forms followed by the New Engineering Contract suite. The contractor is usually content to agree either of these approaches and that ought to be where the contract drafting process ends as the parties settle on with their unamended contract.

Unfortunately, this is rarely the case. The use of standard forms in their unamended state has been described as "rare to the point of non-existence."[11] The amendment may be presented as being necessary to comply with legislative changes but is frequently simply a vehicle whereby risk is transferred onto the contractor. Amendments are proposed and negotiated and then countered and bartered. At this stage, the contractual advisors are less interested in capturing the bargain and more interested in off-loading risk and making a safety net of indemnities against anything that can go wrong. On the other side, the contractor may seek as many caveats and as much room for manoeuvre and argument as possible.

Professor Hibberd makes the point that it is wrong at this stage to blame the standard forms themselves for the problems experienced in the industry. It is the users that make them adversarial either through their "behavioural characteristics" or the wrong choice of contract. It is user-based amendments to standard forms that often become the subject of the disputes.

The result is that there are often discrepancies in the contract-formation process leading to ambiguous and sometimes contradictory contract clauses leading to claims and disputes. MTH v Eon[12] was a typical example of a construction project. It was a case *"with contractual documents of multiple authorship, which contain much loose wording."*

There is a chicken and egg question here as to whether the legal arrangements cause the lack of trust or are a consequence of it. Trust is the cornerstone of commerce as underpinned by law. Trust is an odd term given that the stock in trade is a perception of a lack of trust, particularly in the construction industry. The protection offered by law is usually reactionary – remedies are provided after the loss has been incurred subject to the wording of the contractual documents involved. Trust therefore comes in the form of being able to do something about the other parties' non-performance or poor performance. The instantaneous nature of a data-driven smart contract operation supported by a distributed private or public ledger can remove the need for trust in this sense. The immediacy of a rectified solution at the point at which it is discovered can be applied and allow the parties restitution long before a "cure" of dispute resolution is needed.

The trust we place in our professions has always been a double-edged sword for them. Professionals pride themselves on the service they provide for their customers as the custodians of a body of knowledge and expertise that they can apply. Clients welcome this, and are willing to pay for it on the understanding that they can sue and be compensated if the professional fails to fulfil their role properly. Access to professional insurance is therefore another way of defining modern professional/client interaction. The potential should be explored as to how a data-led approach can eliminate human error and the consequent legal actions which result.

When faced with inconsistencies and ambiguity, the courts are left with their rules of interpretation as were utilised in the MTH v Eon case. The deciding factor was that in a contract for the design, supply and installation of foundations for an off shore windfarm, it was not unreasonable to expect the contractor to warrant for the design life of the installation. The court took an iterative process to determine the most commercially viable and sound proposition of the expressed objective intentions of the parties. The tantalising challenge for smart contracts is whether they could achieve a similar outcome by operating their own interpretation rules on their own potentially ambiguous terms. This might involve machine learning where precedents and smart contract libraries are checked for similar instances and decisions. If could also conceivably involve artificial intelligence as an inbuilt arbiter on matters of interpretation or even extraneous disputes. In the shorter term, the question of whether rules of interpretation are themselves codable is enough of a challenge to be going on with.

The latest development in the public sector for standard form contracts is the Framework Alliance Contract (FAC-1). This contract innovation shows promise in delivering results and accommodating technological advances. One of the features of the agreement is that it is standard form agnostic in that it can sit alongside any underlying contract type. This standardisation of approach would need to be shared in the make-up of a smart contract design, which could alleviate the need to individually negotiate or amend the agreements made.

3.3 Divergence in national legal approaches

Different approaches to construction law around the globe are less pronounced now than in the past. Harmonisation and convergence between different legal traditions is discernible. The reasons for this are increasing uniformity in terms of common business practices and the insistence of lenders, insurers and investors on FIDIC contracts. Common features are also promoted through international treaties, economic and political unions and the adoptions of model laws. UNICTRAL[13] (United Nations Commission on International Trade Law) Arbitration Rules are an example of this phenomenon, which governs the procedural conventions around arbitration which were initially adopted in 1976 and subsequently amended. For example, in 2011, Article 1(2) of the rules was changed to no longer require the arbitration agreement to be "in writing."

The two dominant universal forces at work with construction law are freedom to contract and *pacta sunt serva*. Put simply, the parties are free to agree the terms upon which they contract and are bound by that agreement once made. Differences in approach remain however and the freedom to contract is not paramount in some jurisdictions. For example, in the United Arab Emirates certain mandatory provisions set out in the Law of Civil Transactions take precedence over contractual terms. In most jurisdictions, there are forms of statutory or common law interventions together with the application of general principles which act to circumscribe the two principles further.

The classic distinction given is between the civil and common law traditions. Civil law can be described as a "top down" approach where law comes down from above in the form of codes. Common law takes a "bottom up" approach where the law rises from below on a case-by-case basis, allowing general principles of law to be created from the most unlikely sources.[14] The two approaches converge to the extent that statute law has a dominant role in both traditions. Furthermore, interpreting codified law is left to the courts in civil law countries and this importance is on sometimes equivalent to the case based approach for non-statute-based common law principles such as negligence.

What follows is a series of examples of how national laws diverge around common construction law topics. The purpose is to present the challenge facing smart contract writers to be able to bridge the gaps between the jurisdictions involved and to start to appreciate the size of the task. The international law as stated is for illustrative purposes and should not be taken as a detailed statement of the actual laws involved.

3.3.1 Privity of contract

The rights of third parties to a contract are hampered in England and Wales by the privity of contract rule. Only a part to a contract may sue under it even if the contract purports to confer a benefit on them. One of the leading authorities on this doctrine dates from 1833 in the case of Price v Easton[15] where a contract was made for work to be done in exchange for payment to a third party. When the third party attempted to sue for the payment, he was held to be not privy to the contract, and so his claim failed. Civil law countries do not have such a restriction and, the United Kingdom fell into line with Europe when it brought in the Rights of Third Parties Act 1999 to reverse the operation of the privity rule. However, the Act was capable of being opted out of and that is what subsequently occurred in most standard form contracts on the advice of their lawyers who dis-trusted the uncertainty caused by the Act. The latest position on the Act is that it remains little used in construction contracts.

The privity of contract law does not sit well with the data and digital agenda. The more inter-connected the contracts and supply community becomes, the less it appears acceptable to leave a party unpaid based on a centuries old principle clearly out of step with modern business practice. A meaningful repeal of the principle would allow a whole raft of small firms in the industry to flourish under the confidence of being paid. A smart contract would create a more justifiable environment for payment and increase in speed of payments.

3.3.2 Termination for convenience

This is a widely used clause in international construction contracts. This involves unilateral termination by the employer at their own discretion and without the need for any default to be found on the supply side. Public clients may seek the incorporation of this term into their contracts. The potential for the abuse of this provision is apparent – the contractor would do well to seek an indemnity against the exercise of this clause. The issue of whether or not the contractor is able to recover the profit he would have earned on the unperformed work is clearly an issue. In Brazil, where the practice occurs, the employer is likely to be liable for costs to date and loss of profit. In England and Wales a termination by the client at will would allow the contractor to allege repudiatory breach of contract and to recover damages as appropriate. A termination for convenience clause would similarly lead to uncertainties in the smart contract arena.

3.3.3 Pay when paid

At first glance, a "pay when paid" clause appears to be a sensible way for a contractor to operate in terms of ensuring cash flow remains positive. The contractor seeks the agreement of the supply chain that they will wait for their payment until the contractor is paid. The flaw in the approach is quickly discovered when the position of the sub-contractor is considered. The sub-contractor may have executed their contract perfectly and yet remain unpaid indefinitely.

Pay-when-paid clauses are generally adopted and enforceable under the United Arab Emirates (UAE) laws. This practice was one of the easier targets for Sir Michael Latham to outlaw in England and Wales as enshrined in legislation.[16]

3.3.4 Unconditional bond

Conditional bonds, sometimes known as performance bonds, are common in England and Wales. A performance bond requires the contractor to secure a form of guarantee for 10 percent of the contract price, which will pay out to the client in the event of serious default effecting the performance of the contract. This is therefore a type of security arrangement for the client, which will provide some welcome funds if the client must remobilise with another contractor.

The international equivalent of the conditional bond is a very different animal. Of all the construction law practices encountered internationally, unconditional bonds lead to most head scratching in England as to their true purpose. The client may call the bond, typically 5–10 percent of the contract sum, from the contractor at any time prior to practical completion. The reason for the call may not be stated and the call will be valid in the absence of bad faith.[17]

There is every chance that an unconditional bond may be called in the UAE. The client usually makes the demand directly on the surety bank who readily complies. In Brazil, the courts rarely question the reason why a bond was called and require the call to be honoured. Unconditional bonds are also prevalent in Australian projects, typically with a proportionate reduction in value over the construction and defects correction period. The security obligations around a smart contract including cross indemnity which would require some careful wording.

3.3.5 The ability to sub-contract

A contractor usually takes the risk of the performance of the supply chain in return for a free hand in appointing their sub-contractors. A veto arrangement is usually requested in England although rarely used. In other jurisdictions, the client expects more input on the sub-contractors used with a corresponding ability of the sub-contractor to claim payment from the client in the event of non-payment by the contractor.

In Brazil, this can turn to the supplier's favour in that exceptionally the employees of a sub-contractor have the possibility, if the client has direct powers, to request the establishment of a direct relationship of employment with the client.

In India, the sub-contractor has no privity of contract with the client. Exceptionally, the sub-contractor has been able to claim directly against the client without a nomination procedure being used.

In France, the use of sub-contractors is strictly regulated and must be approved individually by the client before the conclusion of any agreement with the contractor. In the public sector, the sub-contractor must be paid directly by the client. In the private sector, the contractor must provide the sub-contractors with a bank guarantee covering all sums due or arrange for direct payment by the client.

In the UAE, there are restrictions on the percentage of the contract works that can be purchased from overseas. There can be client insistence on the placing of contracts with local manufacturers. Saudi Arabian government projects must sub-contract at least 30 percent of the value to wholly owned Saudi sub-contractors unless none exist. These arrangements can be harmonised in an international smart contract approach. This would allow a balance to be drawn between the different regimes.

3.3.6 A duty to warn?

The duty to warn is one of the more problematic legal principles examined. The vexed question is the extent to which a party has to go further than performing its role and check on the work of others. In England, the principle has been partially established through the law of negligence.[18] In the leading case, it was decided that a reasonably competent builder would not have simply taken the engineer's drawings as face value as to the structural nature of a wall. The ensuing collapse of a building was partially attributed to the builder. This does not amount to a firm principle of law and the courts take a case-by-case approach. This is one of the areas of law where smart contracts and building information modelling (BIM) could lead to some interesting developments in the field of collective responsibility possibly including a pro-rate arrangement such as featured in the Framework Alliance Contract (FAC-1) and Project Partnering Contract (PPC2000).

The duty to warn is not applied in certain US states. For example, the Louisiana Design Sufficiency Law appears to provide the contractor with immunity for defects in the work where they are not to its design.

> *No contractor shall be liable for destruction or deterioration of or defects in any work constructed if he constructed the work according to plans furnished to him.*[19]

3.3.7 Time bar

Time bars stipulate the giving of notice of a claim by a certain time. The failure to give the notice can result in the loss of the right to claim. In England, the parties are free to agree these terms. A consideration of the reasonableness of the term may result from the application of Section 11(1) of the Unfair Terms Act 1977 given their construal as exclusion/limitation clauses.

Any attempt to limit liability for harmful acts are prohibited under UAE law. In Qatar, the Civil Code stipulates that contracting parties may not agree upon a prescription period different to that prescribed by law. Brief notification periods may contravene mandatory provisions.

Time bars are commonly used in construction contracts in Australia. The value of these terms is in allowing claims to be investigated promptly and allowing the client to monitor its financial exposure to the contractor. In Germany, the failure to comply with a notification procedure may preclude a claim from being pursued. An operation of a timebar is probably an attractive feature for a smart contrat where are aligned to dealing with claims in a peremptory manner.

3.3.8 Statutory payment regimes

This work has already intoned the mantra that has long been repeated that cash flow is the lifeblood of the building industry.[20] There are several examples of legislative intervention to improve the position of the supplier often with the expressed intention of limiting the number of insolvencies. England and other common law countries have legislation enacting the Latham proposals and provide a statutory payment protection framework. The Contractor has a statutory right to progress payments in Australia under the security for payment legislation. This has been interpreted and implemented in different ways in the Australian states. For instance, in New South Wales,[21] payments are fast tracked to the contractor within 15 days of the claim having been received. Other security for payment devices include the transfer of a debt owed to a sub-contractor from the contractor to the client. Rapid or immediate payment terms would sit well in a smart contract scenario.

3.3.9 Supplier's lien

In England, once materials have been delivered to site and physically incorporated into the works, the ownership transfers to the landowner regardless of the terms of contract and whether or not they have been paid for. This is at odds with the US concept of the "Mechanic's Lien" which allows the supplier to register a charge over the property as security against payment. This lien complicates matters to the extent that the client requires indemnities from the contractor that it will satisfy any lien and the sub-contractor to issue a release or waiver of any lien registered or unregistered upon payment.

In Brazil, it is possible to lodge a constructor's lien over land and property. Australian security for payment legislation includes the right to secure a lien over the site and any "unfixed plant or materials supplied." The lien is extinguished when the unpaid amount is paid to the contractor.

In Ontario, Canada, the Construction Lien Act extends trust claims to the officers, directors and others who are found to be the controlling mind of the contractor. This statutory remedy allows the corporate veil to be pierced and the directors held personally liable to sub-contractors if project funds are not passed on to the right parties. As in the United States, this can lead to owners to require contractors to carry labour and material payment bonds as part of the security documentation for the project.

In France, regularly used sub-contractors have a direct action against the client if the contractor fails to pay within one month of service of a formal notice. The existence of a right for the sub-contractors to maintain an interest in the fruit of the labour or goods supplied is a strong indicator of supplier bias. The risk of the lien is often passed back to the supplier by requiring indemnities and release forms from the contractor. It is possible that smart contract writers may seek to revitalise the laws around liens or insist upon a system of indemnities which would remove the need for the discussion.

3.3.10 Force majeure/frustration

Force majeure is a French term meaning "superior strength" and is well known in international construction law. The concept is used in common law jurisdictions by inclusion in the contract terms where the only equivalent common law practice is frustration. This occurs when the performance of the contract is radically different to what was intended by the parties and the continued performance of the contract, or a part of it, impossible for a period of time. Force majeure allows the parties to be specific about which events constitute grounds for an extension and their consequences.

Force majeure clauses are available and enforceable in the UAE. Force majeure exclusions are permitted in Brazil and India. Force majeure clauses are available and enforceable under Indian Law and appear as a defined term in the contract. The exclusion of liability for force majeure is provided for in France. The risks can be apportioned as in other jurisdictions. Smart contracts and force majeure clauses would require careful drafting to capture the parties' intentions in such situations. All manner of computer glitches and power outages would need to be addressed in smart contract drafting.

3.3.11 Hardship payments/imprevision

Hardship payments or imprevision payments (as they are known in France) arise from civil law traditions. The Institute for the Unification of Private Law (UNIDROIT)[22] principles set out that a party claiming hardship is entitled to a renegotiation of the contract. If agreement cannot be reached then the party claiming hardship can rescind or amend the contract on "just terms." In the common law tradition, such arrangements appear to lead to contractual uncertainty and allow the proverbial coach and horses to be driven through the bargain terms. However, the clause is unlikely to be able to be invoked for anything other than very strong grounds. In this analysis, the hardship has been viewed as being akin to frustration or commercial impracticability under the common law.

In Brazil, the debtor may terminate the contract if the obligation becomes excessively expensive as a result of extraordinary and unforeseeable events which lead to an extreme and disproportionate advantage to the other party. The affected party may request a judge or arbitrator to modify the contract.

The French Civil Code has the theory of imprevision. The meaning of imprevision concerns an unforeseen obstacle, which complicates the realisation of the contract. This can provide the contractor with the right to claims increased costs and/or to terminate the contract. Examples include increasing costs and wages, interventions by public authorities, social unrest and forces of nature. At first glance this would not be something a smart contract would need to address specifically.

3.3.12 Constructive acceleration

Constructive acceleration can be defined as being a refusal by the client to recognise in a timely manner that the contractor has encountered an excusable delay for which he is entitled to an extension of time. As a result, the contractor may feel compelled to accelerate the programme in order to complete by the original completion date.

A contractor accelerating without being instructed may in some jurisdictions be viewed as able to claim. This has been tried as an argument in England but has met with resistance where it is felt that there are sufficient contractual claims that can be made. In the United States, it is regarded as an acceptable head of claim.

In Brazil, if the constructive acceleration increases contractor costs then it can be successful if an extension of time should have been granted. If bad faith is proven on the part of the client then loss of profits might also be claimed. Australian law allows constructive acceleration where contractors can seek to establish that they had been instructed by implication to accelerate in order to maintain the original completion date. The costs incurred may be recoverable. The same principle applies in Singapore where if dates cannot be achieved by reason of the client's action then the contractor is entitled to recover the costs of additional resources necessary to accelerate the works. In France, there are no established rules dealing with constructive acceleration. However, a decision to accelerate in such things as the threat of liquidated damages may succeed for the contractor. A smart contract equivalent would probably require notice before a claim is made.

The conclusion of this review of the different approaches and practices encountered in international construction law is that there should be an appreciation moving forward of the essential similarities and differences and the need to find a smart contract drafting solution which suits all. The size of this challenge should not be underestimated if the purpose is to write smart contracts, which can apply universally.

3.4 Incompatibilities between existing construction law practice and technological advancements

This chapter has already discussed the differences between reasonable skill and care and fitness for purpose tests of liability. The conclusion made was that fitness for purpose would be much easier to "code" in the sense of transforming the legal test into a verifiable fact discernible upon completion of a construction activity.

3.4.1 Practical completion

The binary position relating to the completion of the construction operations is also something that would need to replace the existing predilection for practical completion. The law of practical (sometimes substantial) completion is another candidate where the industry has agreed a work around to manage a less than perfect set of information and contract administration. The purpose of practical completion is to "de-stress" the end of a construction project. The parties should be able to agree that a project is very nearly finished and that the parties can move onto a different footing whilst the defects are identified and remedied during a rectification period. The issuing of a certificate of practical completion usually sees possession reverting to the client and a release of a moiety of retention from the client to the contractor.

The additional granularity of a data-driven approach would render the broadbrush nature of practical completion of far lesser importance. The device might still be used to some extent. However, it is much more likely that the construction and maintenance of a building through its lifecycle will be part and parcel of the same arrangement with warranties and guarantees provided by the installers. This would be apparent if off-site construction became more prevalent. There are already elements of this in the government soft landings approach. The soft landing referred to is to ensure that there is less of a disconnect between building and usage whereby one should cushion the other by removing their mismatches in expectation and use.

The current state of the law around practical completion was revisited by the Court of Appeal in the 2019 case of Mears Limited v Costplan Services Limited.[23] The contract between the parties contained a clause that the individual sizes of the rooms constructed should not more than 3 percent smaller than those planned. Some of the rooms were smaller and the Appellant argued that this material departure was a breach of contract meaning no valid Certificate of Practical Completion could be issued.

The court recited the law on practical completion highlighting that practical completion was due where the works were complete and free from patent defects except those which could be ignored as trifling. The question before them was whether the 3 percent stipulation represented a material breach of contract. The finding was that it was not a material breach and practical completion ought not to have been withheld. Helpful guidance was also given on the meaning of "trifling" pointing out that the longer the list of defects, or "snags" the less likely the word is to apply. The court frowned upon any poor contract administration practice where the practical completion certificate was given with a long accompanying list of patent defects.

Were the existing contracts to make their way into computer code then there may be a degree of headscratching around how to programme the "trifling" distinction. The coders are likely to prefer a situation where they code whether the works are 100 percent complete or they are not. This is a situation where smart contracts can bring the clarity required and end the debate around one of the grey areas of construction law.

3.4.2 Good faith

The law can be said to be on safer ground when it rules on what people should not do – do not breach a contract and do not act negligently or face the consequences (usually damages). The re-emergence in recent times of the duty of good faith in construction law has seen the law partially moved out of its comfort zone and entertain with a more interventionalist position. This renaissance in good faith in England[24] results from partnering-type contracts, which contain aspirational behaviour clauses. For example, clause 10.1 of the New Engineering Contract requires the parties to "act as stated in this contract and in a spirit of mutual trust and co-operation." This is where the limits of the patterning agenda are discovered. Mandating good behaviour and acting against self-interest represent difficulties for the lawmakers.

The lack of any recent tradition of good faith in English Law has hampered the revival of the concept. Efforts to exhort the judges to make the changes required have had limited impact. The problem remains – how to create an exhaustive list of the type of behaviours expected. Efforts continue down this line which include embedding behavioural experts in supply team meetings who form a view on whether the contributions made at those meetings display the right attitudes to problem solving and teamwork. Those made of the "right stuff" are rewarded through financial incentives. Presumably, the inverse is also true.

Good results are claimed for this approach but one is left wondering if a collaborative approach in meetings is followed up with a more self-interested approach later in exchanges. This raises the age-old foe of the collaborative agenda – human nature and self-interest. An industry brought up on "dog eats dog" has found it difficult to switch into a mindset of mutual trust and co-operation. This is particularly true where the very projects which have sought to promote these values themselves end in acrimony and finger pointing. For example, the case of TSG Building Services v South Anglia Housing decided that a right to terminate at will was upheld notwithstanding its seeming incompatibility with clause 1.1 of Term Partnering Contract which required mutual trust, fairness and mutual co-operation.[25] The parties had formed a dispute around payment and were no longer imbued with collaborative-type feelings towards one another.

One of the main benefits of the technologically driven advancements detailed in this book is that it should no longer matter what the intentions of the parties are. The conundrum of whether or not to mandate good behaviour will be supplanted by a granularity that will see performance as the only measure. To an extent best practice and collaboration may become embedded.

3.5 Emerging responses to the changing landscape

This section examines aspects of the law's development in response to the current pace of technological change. The first of these, drone law, provides a mixture of where existing law can be adapted together with new legislation to give a composite answer to cover the situation. The second looks at case law surrounding BIM and how rules in relation to Common Data Environments have been given a measure of clarity by a court decision.

The Legal Statement on Cryptoassets and Smart Contracts ("The Statement") maps out some of the questions remaining including the interface between intellectual property rights, property itself and shared platforms. The last issue to be addressed pushes the boundaries further in looking for clues as to how ownership and liability for machine learning or artificial intelligence might be dealt with.

The statement recognises the potential rewards for the English legal system and UK dispute resolution. The opportunity is to *"prove a popular foundation for the trillions of smart legal contracts that we may then expect to be entered into annually."*[26] The means of delivering this foundation were recognised as being removing the most fundamental legal impediment by legislation and leaving the common law to do the rest. This approach is examined in the following examples.

3.5.1 Drone law

Drones, or to give them their proper title, unmanned aerial vehicles (UAVs), represent an "indispensable business tool"[27] set to deliver billions worth of value to economic sectors including construction, agriculture, insurance and infrastructure inspection.[28] Drones are already in common usage on construction sites notwithstanding misgivings about security, privacy, aerial trespassing and personal responsibility. This is a prime example of a gap which the law needs to fill – the technology is ready and in use and yet the legislative framework appears to lag. Whilst the general public may wish to buy a drone as a child's present, the attraction soon pales when an air passenger catches sight of a drone out of the corner of the eye when approaching take off in an aeroplane. Happily, this concern has been addressed by the Air Navigation (Amendment) Order 2018 which banned drones from a one-kilometre exclusion zone to an airport or airfield. Security concerns have seen this exclusion zone widened to five kilometres in 2019.

The construction industry's interest in drones is easy to fathom. A drone can capture and produce data in the digital format to record project progress and mapping for BIM. Other functions include surveying and monitoring during the use of a building as well as during its construction. A drone has several advantages over a human operative when performing these tasks including speed, capacity and accessing those hard to reach areas by virtue of being able to fly there. Drones can lead to a great reduction in defects given the ability to conduct a thorough investigation of a build prior to hand over.

The specific legal issues that the lawmakers will have to address around drones include privacy and data protection. The latter has received much more attention in the United Kingdom since the introduction of the General Data Protection Regulations (GDPR) 2018. The likelihood of drones capturing images of individuals, employees, visitors, neighbours and passers-by needs to be considered. An existing set of rules apply to close circuit television (CCTV) and these should, by extension, stretch to cover drone footage. Storage through encryption and subsequent disposal together with express and implied consent from the groups of people identified would help satisfy the GDPR rules.

Trespass into neighbouring air space would create an actionable tort and would require a series of licences or consents to be sought. Oversailing licences for cranes may provide a starting point for preparing any such permissions. Other torts relating to drones include personal injury and property damage caused by collision. Liability would appear to rest on the drone operator in the first instance with subsequent actions perhaps lying against the retailer or manufacturer if the drone is viewed as a consumer purchase. Business to business dealings would likely treat drone liability as falling on the drone operator alone in accordance with any contractual arrangement entered into between the organisation receiving the service and the provider.

The concluding thought in relation to the first example, regulating the drone activity, is that existing legal provisions should be able to cope with this challenge in relation to the construction industry. Construction sites are largely regulated environments under the control of the contractor and employer. Provided adequate permissions are sought and liabilities considered in the terms of engagement then the emerging body of regulation ought to provide for safe drone usage without issues being created outside our legal frameworks.

3.5.2 *Common data environments*

The huge increase in computational and data-processing power of information technology systems has led to the availability of data on an unprecedented scale. Additional factors, such as the falling cost of storage and increasingly sophisticated software systems based on data analytics, have driven the need to capture and share data and design information between the project participants in a common data environment (CDE).

It is becoming increasingly anachronistic to think of data residing in a single place, for instance, on one party's hard drive. Cloud storage involved the physical storage of data spanning multiple servers and is typically owned and managed by a hosting company. The cloud storage providers are responsible for keeping the data available and accessible and protecting and running the physical environment. Organisations then buy or lease the storage capacity from the providers. This type of approach is commonly used in construction by what are termed CDEs. These can be thought of as central repositories where construction project information is housed. The most widespread example of CDEs appertain to the use of BIM. A "BIM environment" is not limited to merely the plans but will also include documentation, graphical models and non-graphical assets. All relevant parties typically have access to the CDE and the co-ordinator of the CDE provides codes or passwords to the platform allowing access for those parties entitled to it.

CDEs and the question of access were considered in the case of Trant v Mott Macdonald 2017.[29] The facts of the case were that the Ministry of Defence employed Trant as the Contractor in May 2016 to provide a new £55 m power-generation facility in the Falkland Isles. Trant had engaged Mott MacDonald as their designers at the tender stage for services including BIM and procurement support. Mrs Justice O'Farrell described BIM as:

> The BIM system is building information modelling. It comprises a software system which is intended to assist the design, preparation and integration of differing designs and different disciplines for the purpose of adequate and efficient planning and management of the design and construction process.

The contract between Trant and Mott MacDonald was never signed meaning that the provisions for use of intellectual property was never finalised. The relationship soured culminating in a dispute where Mott Macdonald claimed unpaid fees of over £1.5 million. When this remained unpaid Mott Macdonald denied Trant access to the servers hosting the design data by revoking Trant's passwords which it had previously issued. Trant sought and gained an injunction from the court ordering that it be allowed to re-access the folders which were intended for use by Trant. The judge held that Trant should make a payment into court pending resolution of the wider dispute. A key factor in the decision appeared to be that Trant had previously had the access and this was now withdrawn for commercial bargaining reasons. Having shared the data up until that point, Mott Macdonald were not in a position to assert ownership rights over the data without being challenged on the same. It should be noted that this was a case for interim injunctive (temporary) relief and did not involve a full consideration of the law. In injunction proceedings, the court need only consider whether there is a substantive cause of action and that a later damage claim would not be an adequate remedy. The judge considered that Trant should have access to the shared data in return for a part payment of the fees outstanding into court.

Newer standard form construction contracts such as the NEC4, released in June 2017, have addressed fundamental BIM issues including liability, use of the model, ownership and information requirements. The NEC4 Option X10 was designed to be independent of other BIM legal statements such as the Construction Industry Council Protocol. Parties are well advised to deal with these issues in their appointment at the outset and throughout. A discussion around these provisions and in relation to the new FAC-1 launched in 2018 follows in chapter nine. The basic approach is to promote data transparency and team integration through direct relationships. Contributions to the BIM environment are sought at all stages of the building process from sub-contractors, manufacturers, occupiers, operators, repairers, alteration teams and even demolishers. The inclusion of the latter team indicates how important it is to design buildings in accordance with the principles of the circular economy. One of BIM's advantages is its ability to deliver on the sustainability agenda through meticulous pre-planning for the whole life of the building.

The ability to turn off access to a CDE has ramifications for smart contracts. Exclusion from the payment mechanisms by rescinding access would be a risk for supply chain organisations fearful of the consequences of putting all their trust in a client that they will allow the smart contract to run its course without seeking to exploit their position as the payer. Securing undertakings around the immutability of the smart contracts once initiated would be crucial in this scenario to grant the security required. This is an argument in favour of using a public blockchain approach.

In conclusion of this example, case law, when given the opportunity, can fulfil an important function in signalling a clear direction on discrete points of law and sometimes their wider implications. There is often a built-up demand for a decision on any given topic about which the judiciary are aware and the opportunity to add clarity is sometimes taken. One of the downsides of this approach is that in litigation there is usually a winner, a loser and a large amount of legal fees to pay on both sides. The other downside of relying on case law is the limited opportunity the judge has to seize the initiative and make some pronouncements of more general import than the facts of the case before the court. As noted,

Trant v Mott Macdonald was an interim decision on only one discrete question around the law's development around CDEs.

3.5.3 Crypto-assets as property

Distributed ledger technology (DLT) enables the sharing and updating of records in a distributed and decentralised way. Participants can securely propose, validate and record updates to a synchronised ledger shared across the participants or nodes. This is an essential platform for smart contracts development. The Statement on Cryptoassets and Smart Contracts[30] (The Statement) defined smart contracts as *"cryptographically secured digital representation of value or contractual rights that uses some type of DLT and can be transferred, stored or traded electronically."*

In this field, the law runs into a problem of a larger scale than those contemplated so far. Fundamental questions such as "what is property?" require a rethink in the context of crypto-assets.

Property, and more specifically the ability to borrow against the value of property, is a cornerstone of society and economy. Although notoriously volatile, crypto-currencies such as Bitcoin have a present value. However, it is the non-availability of using Bitcoin as the security for other transactions that represents an obstacle to wider development. Removing this impediment should be a priority of law to close the gap on what people wish to do with their crypto-assets and the ability of the law to facilitate this.

The issue is that the definition of property, as it has developed in the common law, has an essential characteristic of being capable of being possessed i.e. tangibility or it gives rise to a right that can be enforced through the courts. There are categories or properties, which are intangible, most notably intellectual property rights and dignitary rights such as those protected by the torts of libel and slander. There is also a category of property known as the chose in action where property is being passed between two parties such as by a cheque. However, extending the provisions around chose in action would be to create a false picture as the crypto-asset has a different meaning. The ownership of a crypto-asset is not in transit and rests with the proprietor of, for example, the digital wallet in which the bitcoin are evidenced.

The struggle to keep pace with these developments was shown in the case of *Your Response Limited v Datateam Business Media Limited.*[31] The Court of Appeal held that an electronic database was not a form of property capable of possession and that, therefore, it could not be subject to a possessory lien (a form of ownership). Lord Justice Moore-Bick thought that to include the database in the law of property would constitute a massive departure from the existing law and would require a statute to that effect. None has been forthcoming to date notwithstanding the passage of time since the case. Another of the Court of Appeal Judges, Lord Justice Floyd, conceded that the information would give rise to intellectual property rights such as database right and copyright but could not be property as of right. The key passage here was the statement *"Whilst the physical medium and the rights are treated as property, the information itself has never been."*

The statement recognised that uncertainty around the property question was detrimental to business. Therefore, the question was specifically addressed and the answer given that crypto-assets bore all the indicia of property. On its own, the statement had no formal status as a legal precedent. The first judge to ratify the statement to take heed of this was in the case of AA v Persons Unknown[32] where His Honour Judge Bryan held that "cryptoassets such as Bitcoin are property." This was another case for an injunction following a cyber-attack. The applicant was an insurance company that had paid a ransom in Bitcoin on behalf of one of its customers to secure the reinstatement of the insured's systems, which had been hacked and disabled. On this occasion, it was possible to track some of the transferred Bitcoin to a specified address linked to a currency exchange. The injunctive relief was sought to prevent the dissipation of the bitcoin and to identify the wrongdoers. In order to grant the injunction, the judge had to first be satisfied that bitcoin were property.

The finding that crypto-currencies are property under English Law has significant consequences for the legal rules including inheritance, insolvency, fraud, theft and the law of trusts. Developments in these areas are likely to be forthcoming. All of these areas will also impact, directly and indirectly, on the construction industry and its laws.

3.5.4 Machine learning

Machine learning can be defined as the design of a sequence of actions to solve a problem, known as algorithms, which can optimise automatically through experience and with limited or no human intervention. Construction has yet to see many reported instances of machine learning but it is easy to see how it can be linked to BIM to optimise a building's use based on facilities management responding to internet of things sensors reporting on their functionality and use. Construction applications might look at risk management and health and safety breaches through machine learning.

Legal issues arising include the point of whether parties are bound by agreements – possibly in the form of smart contracts – made by autonomous machines. It is in this field, a whole new approach to law is required. Expecting case law or a single Act of Parliament to govern the area seems fanciful. The statement represents a valiant attempt to keep up with developments and to provide the governance required. Further statements will be required to bring order and confidence in this field. A starting point could be around the issue of where the operation of an algorithm of machine learning produces an unintended outcome.

The case of Arnold v Britton[33] is an example. The case did not feature algorithms as such but did involve the operation of a mathematical formula. The case involved a holiday home lease and the contributions to be made by the lessee towards garden maintenance. The formula used to price the contributions had an anomaly in that in that some holiday homes were charged £1,025,004 per year whilst others paid £1,900. Any sympathy on the part of the judge's to rectify this patent error was not examined – the clause was clear and its operation could not be challenged. This invokes the doctrine of *caveat emptor* or let the buyer beware.

One way to solve the problem posed by unintended outcomes around machine learning would be to introduce a new rule of interpretation around obvious error. Currently, the judiciary are only allowed to employ rules of interpretation where there is some ambiguity on which they are required to rule. Businesses are expected to look after themselves and not typically entitled to consumer protection–type legislation. The difficulty can be with smart contract language or code generally that the users did not understand its operation and this is a challenge for smart contract writers. It is not hard to see a similar set of circumstances arising in Arnold v Britton around the operation of a smart contract, whether autonomously created or otherwise. The availability of in built dispute-resolution provisions are therefore very important.

A more ambitious legal solution would be to give the machine its own status in law. If machine learning, or the more advanced artificial intelligence result in authorship of design or other forms of intellectual property then should that not be recognised legally? Corporations are already recognised as persons at law. Could a corporation with a computer system in control own property and perform as a legal personhood? There appears no reason why not and this could already exist. This is to reach the limit of where science fiction can become fact and where the existing legal arrangements based on contract law become stretched to breaking point. The notion of *consensus ad idem* between willing human parties does not currently fit machine interaction.

Ultimately, to address the changing times, a new theory, with different cultural assumptions and normative underpinning is likely to be required and a different backdrop of state enablement and oversight. This involves, according to Margaret Radin,[34] *"the search for an entirely new theory of contract, which as not based on eighteenth-century visions of individual autonomy and not based on the classical liberal justification of state power. We await the emergence of a theory of justification that fits the social, economic and technological conditions of the contemporary world."*

The new theory of smart contract might very well start with the promotion of transactions as the cornerstone rather than the contracts themselves.

3.6 Conclusion

This chapter has addressed some of the challenges for the legal system arising from technological advancement. The Statement has helped but the contention that only small adjustments are needed to existing laws to provide investor confidence appears optimistic and puts one in mind of Ptolemaic epicycles being promoted as solutions to a much more fundamental objection to the adequacy of existing arrangements. The Radin argument around the deformation of contracts appears to be borne out by some of the examples discussed. Nevertheless, the objective to make UK jurisdiction a state-of-the-art foundation for the developments is extremely worthwhile; it should not though be under-estimated. The next section introduces the smart contract as the solution to the limitations and constraints of the construction industry and discusses how it can operate at law.

Notes

1. Lawtech Delivery Panel (2019) *Statement on Cryptoassets and Smart Contracts*, available at: https://technation.io/about-us/lawtech-panel last accessed 21 May 2020.
2. [1932] AC 562.
3. This is examined in section Five – Online Dispute Resolution.
4. MT *Højgaard A/S -v- E.ON* Climate & Renewables UK Robin Rigg East Limited [2017] UKSC 59.
5. There are clear parallels here between the design obligation in BIM and subsequent technologies. The industry has to move towards fitness for purpose as the only logical standard – the thing works or it does not. The legal development therefore needed is a move away from reasonable skill and care which in any event does not have an international equivalent; hence why, the FIDIC contract states fitness for purpose. In the engineering tradition of a contract, this is much more desirable – the power plant has to work perfectly and we cannot have a situation where it is "a bit wrong.".
6. Another challenge for later lawmakers is the uncertainty caused by having learned judges look at a case and come up with different answers. At one level, this is to be celebrated as evidence of judicial discretion and the wide gambit of the law. At another level, uncertainty in outcomes could prove to be an impediment towards the digitally enhanced future.
7. Formerly the House of Lords.
8. International Federation of Consulting Engineers.
9. Hibberd, P. (2004) *The Place of Standard Form Contracts in the 21st Century*, Society of Construction Law Paper 120.
10. Latham, M (1994) *Constructing the Team*. HMSO, London.
11. Practical Law Company Construction Law Country Profiles, Australia.
12. MT *Højgaard A/S -v- E.ON* Climate & Renewables UK Robin Rigg East Limited [2017] UKSC 59.
13. United Nations Commission on International Trade Law.
14. Donoghue v Stevenson [1932] AC 562.
15. Price v Easton (1833) 4 B & Ad 433.
16. Housing, Grants Construction and Regeneration Act 1996 & Local Democracy, Economic Development and Construction Act 2009.
17. Bailey, J. (2016) *Construction Law*, Routledge at page 1074.
18. Edward Lindenberg v Joe Canning and others (1993) 62 BLR 147.
19. Louisiana Revised Statute 9:2771.
20. Danways Limited v F G Minter Limited [1971] 2 All ER 1389.
21. Building and Construction Security of Payment Act 1999.
22. International Institute for the Unification of Private Law.
23. Mears Limited v Costplan Services Limited [2019] EWCA Civ 502.
24. Mason, J. (2007) *Contracting in Good Faith-Giving the Parties What They Want*, Construction Law Journal 23(6) 436–443.
25. TSG Building Services v South Anglia Housing [2013] EWHC 1151 TCC [2013] EWHC 1151 (TCC).
26. Lord Justice Vos, *Crypto-Assets as Property: How Can English Law Boost the Confidence of Would-be Parties to Smart Legal Contracts*, Lecture in Liverpool, U.K. May 2019.

27. Winfield, M. (2018) *Drones: The Legal Frontier – The Opportunities and Risks of Taking Flight*, Society of Construction Law Papers D213.
28. In 2016 Price Waterhouse Cooper estimated the market value of drones in commercial applications as being over $127 billion.
29. Trant v Mott Macdonald [2017] EWHC 2061.
30. Lawtech Delivery Panel (2019) *Statement on Cryptoassets and Smart Contracts*, available at: https://technation.io/about-us/lawtech-panel last accessed 21 May 2020.
31. *Your Response Limited v Datateam Business Media Limited.* [2014] EWCA Civ 281.
32. AA v Persons Unknown [2019] EWHC 3556.
33. Arnold v Britton [2015] UKSC 36.
34. Radin, M. (2017) *The Deformation of Contract in the Information Society*, Oxford Journal of Legal Studies 37(3) 505–533.

Section III
TO BE

4 The smart contract in construction

Yogi Berra once said, *"it's tough to make predictions, especially about the future"*[1] This joking truism counsels caution and a wait and see approach to predicted developments. This is particularly sage advice in the fast-paced world of technological innovation. However, part of the role of an academic is to future gaze and help to prepare the ground. The possibilities for new approaches to be adopted in the construction and engineering sectors currently feel limitless. Consequently, these are interesting times. This is partially because many of the innovations and initiatives have already made substantial progress in other industries; take, for example, robotics in vehicle manufacture. The leaders of today and tomorrow should therefore be exhorted to look to windward and set a course to deliver the improvements from which the construction industry has been missing out. A clear path now appears within reach through the medium of smart contracts and enhanced collaboration.

In 1965, the founder of Intel, Gordon Moore, predicted that every two years we would be able to double the number of transistors put on a computer chip. This theory has largely been proved and is still going strong and is predicted to hold true for decades. This is exponential or explosive growth. This is not just limited to processing power. Other technologies – hard disk capacity, internet traffic, bandwidth, magnetic data storage and random access memory – are also growing at similar rates.[2] This brute force processing power stored on a "cloud" offers seemingly limitless storage capacity, lightning quick communications, ever-greater miniaturisation and rapid decline in the cost of components. Smart contracts are one way of harnessing this massive potential.

Contracts are the currency of commerce. Insight into the future of commerce is likely to be gained through the application of smart contracts. The improvements in technology and the direction of travel for the collaborative agenda seem to coalescence around the concept of the smart contract. This vehicle can ultimately give an application for the distributed ledger technology (DLT) and crypto-currency about which so much noise has been generated in recent times. Here too, is an embodiment of collaboration – people trusting each other on a shared platform. This "To Be" section therefore explores the huge potential of the smart contract to be beneficial for the construction industry and to address some of its limitations discussed in chapter three. The legal infrastructure needs to be there to allow them to flourish. Opinions differ on the extent to which existing law can facilitate their adoption.

Lawyers and legal academics have become increasingly aware of the challenges and potential of smart contracts in recent years. Some commentators view them as business as usual[3] whilst others see the end of contract law, as we know it.[4] Movements also exist to do away with existing national laws and to start afresh in a system free of the traditional constraints and hierarchies. The author's own view is that smart contracts can represent the paradigm shift previously discussed as the "helio-centric moment". However, interaction with existing law is required, and is perfectly feasible, in order to harness the benefits on offer. One of the main advantages the new viewpoint will convey is the ability to rationalise and remove the ambiguities in the existing contract law arrangements.

Mentioning smart contracts to a construction professional is likely to draw from them notions around the automation of construction performance. The emphasis here is on how the smart contract will work in practice. From the lawyer's perspective, this is not the challenge so much as how the contractual language can be formalised into being legally robust whilst being machine-readable. This is the key aspect of the definition of smart contracts.[5]

The term "smart contract" was first introduced by Nick Szabo in the early 1990s.[6] However, it was only with the advent of blockchain and DLT that this concept gained widespread interest.

Szabo's revolutionary idea was to embed contractual clauses in a digital entity (later to become DLT) with control over the property involved. The digital entity would use secure, machine-executable transaction protocols ensuring automatic performance of predefined, conditional actions in accordance with the contract clauses. Market demand and technology were not ready at that time to support this approach. They are now ready and able to provide the platform and to go further than Szabo's original ideas.

The appeal of the smart contract is in its simplicity. Transaction costs and times are reduced by digitisation, the end users need not trouble themselves with the internal workings and risk allocation comes as standard. This standardisation is a result of the economies of scale involved in smart contracts. The contracts are reduced to simple earned value transactions and the terms become non-contentious in the pursuit of this basic formula. Having agreed the base lines, there is no need for bespoke amendments. This can be characterised as thousands of mini-contracts leading to an inch-stone progression towards completion. Each mini-contract is independent of the others and has clear functionality and execution. The extent of the automation may differ but even small steps towards this end offer huge benefits in time, cost and quality.

The potential is there to automate the complexity and leave the users with a straightforward transaction-based interaction, which is called "earned value" here. Party A pays Party B for the services and goods received based on the value generated to Party A. There is no need to refer the transaction to any other wider context or value calculation. This is a back to basics merely transactional approach suitable for common adoption insofar as the threshold for appreciating the functionality is much reduced and Hibberd's goal of understanding is met. The latter will always be found wanting where complexity clouds understanding and judgment.

The idea that complexity should be automated where possible is replicated in popular apps such as Uber and Air BnB. Where everything is taken care of in the background, the participants are left to interact in the foreground in a mutually understood and convenient fashion. The passenger knows that the driver will receive a fair price and do not need to glance at the meter nervously as we sit in traffic. The host knows that the guests have been pre-vetted and have loyalty points and good reviews. The interaction is bound to be much more easily forthcoming and be pre-programmed to have trust and confidence in each other. This would render redundant the need for precedent as per Hibberd's goals.

The key characteristics of smart contracts is it in digital form and is embedded as code in hardware and software. The performance of the contract and the release of payments and other actions are enabled by technology and rules-based operations. Lastly, the smart contract is irrevocable as once initiated, the outcomes for which a smart contract is encoded to perform cannot typically be stopped. Performance is automatic through a network of data sensors and automated ledgers. The quality-checking function can be augmented through using technologies currently being developed. In the short-to-medium term, this is likely to require continuing human involvement. Smart contracts probably require a building information modelling (BIM)-type model on which to base its assumptions as to the fulfilment of the planned versus actual performance. Another potential route for development is to be independent of BIM and take an app-type approach to smart contracts. This is a semi-automated position where the certifier takes images of the work and materials for checking. The checking is performed automatically together with the cursory manual inspection of the priority areas.

4.1 The component parts

Smart contracts require the following as a minimum in order to function: BIM level 3 and beyond, crypto-currencies and the blockchain, big data/internet of things, and appropriate payment mechanisms and liability arrangements. A picture is worth a thousand words, so the saying goes, and Figure 4.1 has sought to convey the sense of smart contracts in construction to many different classes of students. The explanation starts with the picture and then breaks this down to the component parts as set out below.

The smart contract process can be described thus: the operative (whether human or robotic) inserts the brick in the wall. The presence of the brick is recorded by the sensor. The quality of the installation is checked against the desired criteria. The presence of the warranty information is verified and payment is released to the installer/supplier. The transactions can be recorded on a

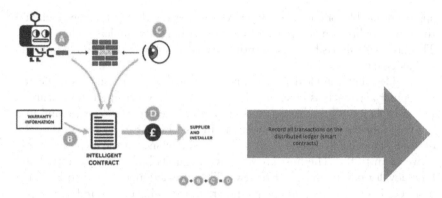

Figure 4.1 The components of a smart contract

Figure 4.2 Robot placing brick in the wall

distributed ledger using blockchain technology. This can involve a brick-by-brick certification of completion if required – an "inch-stone" approach to building rather than the traditional milestone (Figure 4.2).

The key characteristic of smart contracts is the coding of legal terms and processes into software. The result is that any contractual response is the outcome of a "rules-based" technological operation. Contractual responses are automatic once initiated and typically cannot be stopped (immutability) or reversed once commenced (Figure 4.3).

Figure 4.3 IOT sensors pick up and record the brick's presence

WARRANTY INFORMATION

Figure 4.4 Necessary documentation

The "sensing" of the brick being in place is not a massive challenge for the construction industry. The insertion of the brick can be monitored by big data sensors or observed by an oracle. The challenge is in recording the work has been completed on the same platform as the smart contract operates. The brick placement must be sensed and recorded and its presence uploaded onto a DLT from, which the smart contract looks to for its single source of truth. The smart contract is not in itself sentient in that it does not know when the brick is in place. It must be told that this is the case via the ledger. The ledger can therefore be described as a specific configuration of technology components that records and tracks information in a distributed (as opposed to centralised) manner (Figure 4.4).

The importance of any work performed to be supported by warranty or guarantee is likely to become of increasing importance. The current limitations of the existing law are seen here in how the industry clings to standards of reasonable skill and care rather than fitness for purpose, which is required to give a cast-iron warranty for the performance of the component. The discussion around whether this would be a design or performance warranty is yet to be resolved. Contractual safeguards such as the need lodging/checking of the warranty information could be provided through the blockchain as well as the record of the financial transactions taking place.

Oracles for smart contracts are a form of disaggregated professional input (Figure 4.5). Oracles could also input the "reasonable" judgment – an example of exception handling, in computer speak. Central to the ability for the contract to be automated is the monitoring of performance. A system is required whereby the monitoring of external inputs from trusted sources can be undertaken in order to verify in accordance with contractual stipulations. Spotting the defects or dealing with extraneous events – the overseer has been deemed the oracle. The

Figure 4.5 The oracle sees

Figure 4.6 The currency is paid

term oracle is un-necessarily mysterious. The original oracles of ancient Greece were able to make predictions and communicate with the Gods. The eye which observes the installation of the brick can be machine or it can be human. The various oracles on a building project might be a hand-held smart phone, temperature gauge, sensor attached to a steel girder, drone or observation by a contract administrator in a site hut. Each and every one of these are capable of creating data on the status of the build for uploading to the DLT.

On the model, the pound sterling symbol is used to represent the consideration for the delivery of the component parts. However, crypto-currencies may in time replace ordinary currencies and their accompanying banking arrangements. One benefit of this approach would be to take any restrictions of how often payments were made and the frequency with which one went to the bank. Valuations and payments could become real-time "inch-stone" payment rather than milestone or monthly valuations in arrears (Figure 4.6). The development of recognising crypto-currencies as property in the case of AA v Persons Unknown[7] has prepared the ground for further uptake.

Smart contracts can also make considerable improvements in securing payments for the benefit of the supply chain and are an opportunity to protect parties from insolvencies and late payments. Digital currency presents the opportunity to settle invoices much more quickly and therefore to distribute payments much more quickly across the value chain and to avoid paying interest and bank charges that ultimately hurt the value chain itself.

Handling payment within the system on a distributed ledger can guarantee the integrity of payment and, using the features inherent to distributed ledgers, can encourage good behaviour through financial incentives (Figure 4.7).

Figure 4.7 The record on the blockchain

4.2 The distributed ledger

A ledger in the traditional sense is a book or other collection of financial accounts. Double entry book keeping has been used in accounting since Roman times as a form of ledger. Every entry to an account requires a corresponding opposite entry to a different account. One is known as a debit and the other as a credit. Keeping the ledgers in balance ensures there is little room for accounting error. The same principle applies to DLT save that there are not simply two books but potentially hundreds of thousands of the same ledgers distributed through the nodes of a blockchain network. The popularity and potential for blockchain approaches are partially explained because in this approach, the room for accounting error in this scenario is reduced to zero.

The crucial development in blockchain and crypto-currencies happened in 2008 when the pseudonymous Satoshi Nakamoto published a white paper entitled "Bitcoin: A Peer-to-Peer Electronic Cash System."[8] This key event provided the groundwork for the distributed ledgers on which smart contracts operate. The natural tendency is to associate blockchain with crypto-currency and conjure images of a lawless wild west-style volatile commodity. However, crypto-currencies represent only one of a thousand uses for blockchain that includes uses as varied as tracking shipping containers around the world and a system of food provenance allowing perishables to be traced back to their farm of origin. The complexity of a building project would not phase such a sophisticated and vast resource as blockchain technology.

Therefore, it was not until DLTs became available that Nick Szabo's ideas could come to fruition. This is recognised by the writers of the recent International Standards[9] where the ledger is clearly seen as the most important part of the smart contract set up. Their definition of smart contracts specifically references the DLT: *"a computer program stored in a distributed ledger system wherein the outcome of any execution of the program is recorded on the distributed ledger."*

The ledger systems used can vary from private off-line ledgers – a list of peers or stakeholders on a project – to a public online ledger rewarding a random data miner for their processing of the transaction. The ledger operates a consensus mechanism (or algorithm) by which activities are verified and entered onto the ledger.

When all of these components are put together on a building site, the result can be this:

Figure 4.8 shows the integrated nature of the smart contract approach. The performance of the smart contract obligations is recorded by the oracle/collaboration sensors and is then transformed into a binary code on the ledger. The information is then fed back to the client who may check the progress against planned progress and fulfilment of tasks on a dashboard they operate for the build process. It is straightforward to imagine that the BIM design can also access the same technology and give the client real-time data on the build down to a considerable level of detail based on the data sensors.

The component parts of the smart contract have been broadly outlined here. It is now necessary to take a wider view of their benefits and potential usage along

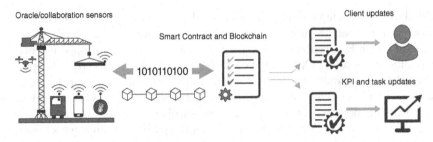

Figure 4.8 The smart contract from the ICE blockchain guide[10]

with how they can be deployed within the constraints and particular habitat of the construction industry.

4.3 The benefits of smart contracts

The proceeding chapters have been making the case for smart contracts as the logical progression from existing arrangements and focal point for technological development. The case is made here for their adoption in terms of the benefits they bring.

4.3.1 Reducing transaction cost in business relationships by making third-party services obsolete

The traditional banking system whereby individuals and corporations hand over their money to a third party for safekeeping is under threat from a distributed ledger approach twinned with a crypto-currency. The disintermediation of this role (see chapter 1.3) would come as a benefit to those users currently paying for the services. It is quite possible for the role of bank to be re-intermediated to provide another or related service is likely given the evolution of commerce. People will remain extremely busy and paying for a service of currency handling is unlikely to become redundant any time soon. However, smart contracts do not need third parties in the same way as existing payment arrangements happen on building projects. This means that once the precedents are created and the modus operandi is settled, there would be considerable savings in transaction cost and the delay taken by moving money around.

4.3.2 Reducing risk

Risk allocation in the construction industry and the contract draftsmanship designed to apportion risk leads to distorted risk balances and liabilities being placed on the unwary or blasé parties. Unamended contracts are extremely rare in the construction industry as the client seeks the optimum risk profile.

Smart contracts have the potential to massively increase the granularity on a construction project adopting an inch-stone approach whereby risks can be micro-managed and appropriate adjustments taken during the planning and construction phase itself. Once established, a standard risk template would be adopted and there should be no requirement to amend the same save to particularise the smart contract for the site in question.

4.3.3 Significantly reduce room for misinterpretation and misunderstandings

Another by-product of the granularity of a smart contract is that this allows much more precise parameters to be set for contract provision. Concepts such as practical completion and the decision around whether or not a defect is "trifling" would hopefully be consigned to history. If an error does occur in the code then the dispute avoidance or resolution process used to correct it should have a linear solution – this went wrong because of this. The known effects were recognised at a much earlier stage and it is rectified that much more straightforwardly than in an after the event dispute resolution procedure at huge cost and time.

4.3.4 Accelerated processes

Smart contracts are part of a wider movement of technological innovations that includes off-site manufacture and modern methods of construction. The Institute of Civil Engineers (ICE) blockchain report (Figure 4.9) gives an example of a steel beam identified by its unique code reference, which accompanies it from manufacture through transportation to site. The ownership and insurance requirements in relation to the steel beam may change during its journey from fabrication to site and these smart contractual arrangements can themselves be recorded on a distributed ledger. "Just in time" delivery and lean construction have been around for some time. The additional detail of where the steel beam is, its exact dimensions and state of ownership would allow for much quicker processes around construction management and execution.

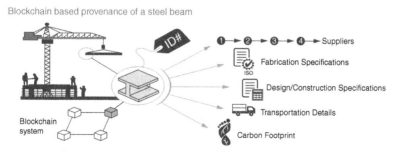

Figure 4.9 Provenance of a steel girder (attributed to ICE blockchain publication)

4.3.5 Standardisation and e-procurement

Considerable value can be extracted from smart contract precedents, which explains the current clamour around seeking to create the language and domain servers selected for use. It is not difficult to envisage a standard form suite of smart contracts capable of delivering a project based on user experience and refinement. This is the goal of initiatives like the Accord Project, which aims to create a smart contract language understandable by lawyers, computer scientists and the end user. These smart contract precedents would cover all aspects of construction from manufacture through the suppliers and include all tiers of the supply chain.

4.3.6 Transparency and traceability

The symbiotic relationship with distributed ledgers mean that all parties involved see the same code at all times and its current state of execution. The system is constantly backed up and rendered secure by its distributed nature. The consensus mechanisms provide each participant with unequivocal evidence of the occurrence of relevant events and the results obtained. Add to this the certainty provided by the immutability of the smart contract once initiated and there is a compelling case for supply-side and client-side confidence in the transactions and procedures implemented.

4.4 Smart contract development

The ISO (International Organisation for Standardization) is a worldwide federation of national standard bodies. Their role is to develop high-quality voluntary international standards which facilitate international exchange of goods and services, support sustainable and equitable economic growth, promote innovation and protect health, safety and the environment.[11]

Crucially, for present purposes, the ISO is a forerunner of actual lawmaking and paves the way for smart contract adoption it their 2019 British Standard Institute publication *"Blockchain and Distributed Ledger Technologies-Overview of and interactions between smart contracts in blockchain and distributed ledger technology systems."* Reassurance and guidance are established first through the medium of a standard approach which this document seeks to usher in. The standard highlights the different approaches to smart contract development.

4.4.1 Szabo's approach

The term "smart contract" was coined in 1994 by Nick Szabo, a cryptographer, who defined it as *"A computerised transaction protocol that executes the terms of a contract. The general objectives of Smart Contract design are to satisfy common contractual conditions (such as payments terms), minimise expectations and minimise the need for trusted intermediaries."* The effect of such contracts on contract law and economics, and their opportunities were said by their originator to be *"vast but little explored."*

The actual approach here was not to automate contracts, but instead to automate contractual clauses (the stack approach). Szabo's idea was more about automating the evidence and consequences of contractual agreements and not the contract itself – particularly, its creation and enforcement. In his paper, JG Allen supports the Szabo approach and considers that most smart contracts will remain "stacked" in a more conventional contractual framework that governs the parties' relationship around the software processes rendering performance. These "stack" contracts will remain nestled in a national contract law regime. It is the latter, which can deal with legality, translation and ambiguity. The role of a human "semantic oracle" could sit alongside other oracles to decide issues relating to the conditions of contract performance, particularly where circumstances have changed in unpredictable ways.[12]

Szabo drew an analogy with a vending machine where the payment has to be received before the fizzy drink is dispensed. This *"money first, goods second"* approach can be disregarded if the technology allows for the seamless, real-time exchanging of goods/services for money. The use of big data and censors allows payment to be made instantaneously through crypto-currency.

Other definitions have sought to extend the notion that only certain contractual conditions are automatable.

4.4.2 The Lessing approach

"Code is [shall be] Law."[13] This arresting pronouncement presents itself as a kind of Elderado to intrepid travellers seeking to write a smart contract at once a legal document and computer programme. This is widely seen as an over-simplification because although both computer code and law are deterministic; this glosses over issues such as enforcement and jurisdiction. However, it is possible to say that "The code is the smart contract" and that Szabo's limitations on the scope of smart contracts can be removed.

An extension of this approach is to predict contracts that are fully executable without human intervention. These contracs can be self-enforcing, monitoring external inputs from trusted sources in order to settle contractual obligations according to the contracts stipulations'. The application of smart contracts in the banking and investment sectors appears easier to establish given the relative straightforward nature of the instruments involved. The general rule is that the longer the contract, the less straightforward its automation;[14] therefore, a semi-automated position appears to be the likely outcome in the short-and-medium term. However, the infrastructure is now available to allow smart contracts to be written in fully fledged programming languages, to communicate and interact with each other as well as with external resources, and to transparently keep track of their current state of execution.

4.4.3 The lawyer's approach

The lawyer's approach is to take a much more pragmatic view of smart contract development. This is based on the universal principle that legal validity cannot be denied unless applicable jurisdictions explicitly require another form of performing certain transactions.

The simple fact is that the legal character of a smart contract is whatever a judge or the law decides it to be. There is a reminder here to the would-be programmers of smart contracts of who is in charge when it comes to making and interpreting the law. The crux is to remember that a contract is not a statement of law in itself – it is an agreement between parties that needs to comply with contract law.

The take away from these different approaches is that notwithstanding the novel aspects of a smart contract it still needs to fall line in terms of its legal status.

4.4.4 Language as the key

A smart contract binds natural language text to computable code via a data model. The benefit is to deliver automation, or at least semi-automation. Taking the components in turn:

1 Natural language text – Legal systems are based entirely on the contract wording to deliver certainty and allow our statutory and common/civil law regimes something to fix on. We, as humans, need to trust and understand the undertakings being exchanged. Ultimately, the words will not be needed as the focus moves from the contract to the transaction itself.
2 Computable code – Computer code and law have many similarities. Both express "logic" and need to be understood to be of use. One functional difference is that one is readable by a lawyer and one by a programmer. The Accord Project, which held its inaugural conference in the City of London in June 2019, has sought to establish a language and a set of templates that are readable by both.
3 The **data** model – This can be planned and actual as an internet of things approach updates the model in real time. Essentially, each component of the build becomes its own mini-contract (think inch-stone rather than milestone). The natural starting point for data programming and extraction is the BIM software. This requires construction clients to appreciate that the value of the model is not only in the construction phase, but, much more so in the facilities-management phase. The model can become the nerve centre for the building's performance and contribute to making it more sustainable in its use and maintenance.

The net effect of having these three components in one artefact provides a framework for constructing a building or civil engineering project. The sheer volume of mini-contracts being executed and performed leads to the involvement of a blockchain or similar DLT.

The steps along the way towards the development of a smart contract language were considered in Figure 4.10.

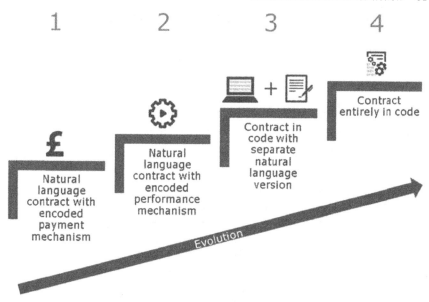

Figure 4.10 The development of smart contract language (Ashurst[15])

The key concept at every stage of this development is the language itself. The stack approach is discernible in first automating parts of the contractual obligations alongside a natural language before taking the steps towards the contract entirely in code. The interim measures appear as encoding first payment and then performance before a dual existence before giving way to the contract becoming fully written in code alone. At this last stage, the goal of having a language readable by computer scientists and lawyers will have been realised. Leading American universities are already teaching their law students to code so that they can contribute to this movement.

Natural languages such as English have evolved from pre-history to the present day. The language of the law is a subset of a natural language. Formal computer languages are much more recent and have been written to allow communication between human and machines using binary logic. It is tempting to follow the Lessing approach of Code is Law and Law is Code. However, there are important differences despite certain similarities in terms of structure and symbolic content. One of the most important of which is dealing with situational sense. A good example of this comes from the case of R v Bentley[16] where two accomplices, one with a gun are confronted by a police officer. *"Let him have it,"* are the infamous words uttered by Bentley who meant either to shoot the police officer or hand over the gun to him. It was difficult enough for the Judge to take a view on the meaning on the facts of a case let alone imagine how a computer could be programmed to come up with the correct answer. In the event, Bentley was hanged for murder, one of the last instances of the death penalty in the United Kingdom. Having situational sense and dealing with extraneous events will be a major challenge for smart contracts in construction and is likely to prevent full automation in the interim period.

Taking this point to its extreme, if an interaction with the wider world requiring the support of trusted entities cannot be dispensed with, then the smart contract ambition – to do away with third parties in a decentralised system – has been compromised. Nevertheless, and in common with other unicorns such as digital twins – it is the progress towards the full outcome that represents achievement and savings in time and cost on the present arrangements.

4.5 The features of a smart contract

A smart contract is a piece of code that is stored on a ledger, triggered by ledger transactions and which reads and writes data in that ledger's database.

The blockchain can be thought of in terms of a tube train (probably driverless) travelling between stations where some passengers alight and others leave the train. The addition of passengers is like adding data to the blockchain. A passenger who stays on the train for a couple of stops and then alights is analogous to a smart contract – the consideration has been fulfilled in that the passenger had been transported in return for the price of the ticket.

The interaction between smart contracts and distributed ledgers establishes the key features of a smart contract, namely:

- Automated – the execution of smart contract code requires no human intervention
- Deterministic – given the same initial conditions, the executed code should always give the same result
- Virtual – both the conditions and the consequences of the contract require no formal document and must be represented within the ledger
- Unalterable (immutable)– the conditions and consequences of the contract cannot be altered, except in ways that were originally anticipated within the contract itself
- Irreversible – transactions that occur in accordance with the terms of the contract create permanent changes within the ledger
- Available – anyone (who has permission) is able to trigger the execution of the contract code at any time
- Transparent and auditable – the code, input and output of the smart contract is reviewable by any member of the network

The above features could ensure that the complexity of construction agreements could be tackled by the automated processes taking place within distributed ledger systems. For example, a business could programme a smart contract so that when a component is installed in a building during construction, the smart contract is triggered, the supplier is automatically informed and currency is released. This explicit interaction with a distributed ledger means that changes are triggered on the same ledger, as well as being stored on the ledger. The executable code (the smart contract) itself has the same properties as other records on the ledger.

4.6 The relationship with distributed ledgers

The inch-stone approach of smart contracts has been identified as only working where there is sufficient data generated around the execution and completion of the smart contract and the transactions are recorded. This is where the internet of things and the blockchain comes in. This technology which blurs the lines between the physical and digital spheres is described as the fourth industrial revolution. The first industrial revolution was steam power (1760–1840), the second was electricity (1870 to 1914) and the third involves the use of micro-processors (1970).

A blockchain is a ledger, or a database of transactions recorded by a network of computers. Often referred to as the DLT, transactions are grouped in blocks and the chain forms the history of these transactions.

The blockchain appears to be a much more stable and trusted platform which has gained the interest of the global corporations. The analogy here is of a driverless tube train that stops in the station at exactly the designated place. The doors open and the transactions either get on or get off. The doors close and the train moves onto the next station. This is the logic of the blockchain. The train represents a huge data string of carriages which can be added to infinitely. The cross verification of the process by multiple reference points prevents the abuse of the system.

Taken to the construction context, it is easy to see how the interim payment for component parts of a build could use blockchain technology. Each component is individually chipped and once big data sensors attest to its successful installation and function then the payment will be made at the next blockchain station. Human intervention here is not strictly required. The simpler the construction or engineering component being undertaken, the better in the first instance. Laying rails or achieving electrification of a line could be relatively simply ascertained. More complex build items may present more of a challenge but not an insurmountable one.

One argument seen is that the binary nature of the blockchain exchange is inadequate for financial instruments and security payments. The funds can be released upon electronic execution of the bond or warranty documentation. The argument is that in more variable arrangements the computer will not be able to cope.

In September 2015, the World Economic Forum listed bitcoin and the blockchain as one of its technology tipping points, expecting that 10 percent of the world's gross domestic product (GDP) will be stored on blockchain technology by 2027. It is estimated that the first taxation will be collected by the government via a blockchain by 2025.

The key properties of a DLT that enable them to inform trust in transactions are:

* Shared read – blockchains are a structured data store that many people can read
* Shared write – as well as read, many people can write data into the database
* Absence of trust – the different writers do not have to trust each other not to manipulate the shared database state

- Disintermediation – there is no need for a trusted intermediary to enforce access control
- Transaction interaction – records in the database depend on, and link to, each other
- Validation rules – rules around database transactions are well defined, such that anyone with a copy of the database can validate that it had been maintained correctly

The situation arrived at is a platform for business free of hierarchy and "trustless" in the sense that it is self-policing, simple and fair.

In any discussion about distributed ledgers, it is necessary to make a distinction between public (on-chain) and private (off-chain) ledgers. In either case, digital signatures are an important tool in convincing validators of the veracity of real-world data.

An on-chain transaction, simply called a transaction, occurs and is considered valid when the blockchain is modified to reflect the transaction on the public ledger. It involves the transaction being validated and authenticated by a suitable number of participants, recording of the details of the transaction on the suitable block, and broadcasting of the necessary information to the whole blockchain network, which makes it irreversible.

An off-chain transaction refers to those transactions occurring on a network, which move the value outside of the ledger. The off-chain transactions may eventually be recorded on-chain. The suitability of an off-chain approach for construction is demonstrable when consideration is given to big data sensors which are operating local software and the other data creating entities

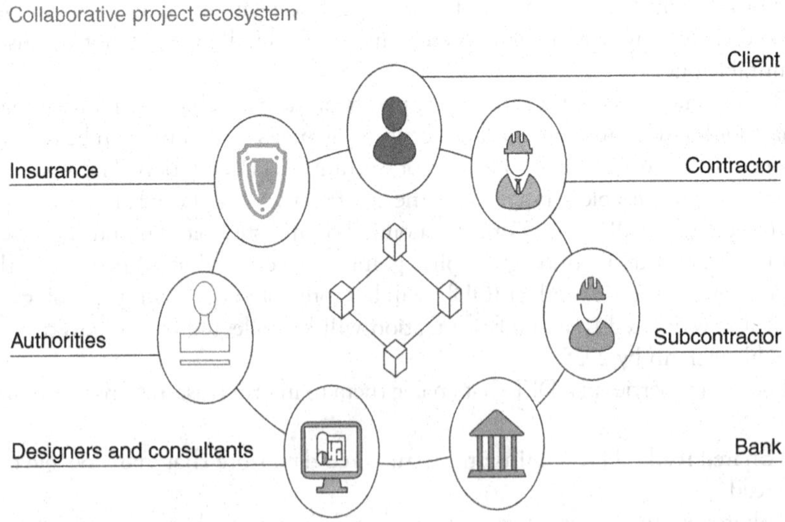

Collaborative project ecosystem

Figure 4.11 Collaborative project ecosystems (attributed to ICE blockchain for construction)

such as websites providing information such as weather services, news or similar. Public systems using miners who randomly change cannot be attached to the sensors as they rely on the blockchain for their transactional data. For their purposes, the data has to come from an oracle, which is able to connect to a blockchain.

However, where the big data sensors are connected to transactions operated in an off-chain network then the communication and ledger entry of the data can be achieved. The risk of inconsistent data from the big data sensors being generated by having multiple sensors or data points for values can be offset by conducting plausibility checks over multiple sensors and harmonisation values. These off-chain transactions can be recorded on a ledger at appropriate intervals. The ledger used for this is more likely to be a private one, although could conceivably be public at this stage.

4.7 An example of a private off-chain ledger

The further benefits of off-chain transactions are:

- Scalability – Public online ledgers are not currently suited to managing large throughput. This can be problematic where you are sharing a network with other applications. Running smart contracts off-chain means that a dedicated server can be used for the project transactions.
- Instant execution – On-chain transactions have a lengthy lag time depending upon the network load and number of transactions waiting in the queue to be confirmed.
- Privacy and security – Computation occurs off-chain and details are not publicly broadcast. In some instances of on-chain transactions, it is possible to derive parties' identities by studying transaction patterns.
- Limited ability to rectify mistakes – If the private ledger has to be changed, then the simplest way may be an inverse transaction to a previous transaction. This will not undo or delete the respective entry but will eliminate the cumulative outcomes.[17]
- Cost – Computation cost for on-chain transactions can become high and volatile, particularly, if linked to crypto-currencies. Furthermore, off-chain transactions do not require "data mining" making this an attractive option especially where large amounts of transactions are involved.

One area where the stakeholders in a private ledger may retain some nervousness is around the issue of control of the system which the client retains. The risk is on a private ledger that the client is still ultimately in control of the flow of the consideration and therefore the playing field remains levelled in their favour. Applications where software as a service is used could be the answer here. A third party with duties towards the project rather than the client could administer the smart contract process and maintain the ledgers independently. However, all databases including distributed ledgers require digital storage and processing power to carry out operations. In a private setting, this comes from the project budget. The key question is how to pay for the operation of the system in a way that retains trust of all members of that system. The client might pay for someone to design and set up a private or permissioned system which the others can join at relatively low cost. The phenomena of an "unshackled oracle" could provide some comfort to the supply chain that once initiated, the smart contracts will run through to payment. This model could be based on the project bank account, which is discussed in chapter 9.3. However, even this seeming magic bullet to the problems of payment on construction project still requires the funds to be placed in the bank account in the first instance.

Comparing the two routes for smart contract completion set out in the International Standard 23455:2019. Smart contracts usually operate in public online systems following a three-step process: deploy-invoke-operate.

- Deployment – A smart contract is made available to the blockchain system with a transaction containing the pre-compiled data. In the deployment process, this code transaction obtains a smart contract address for later invocation or operation. The code itself is not active yet. Once it is added to the distributed ledger, it is practically immutable; however, later on, it may be cancelled or replaced with an upgraded version just like a business offer.
- Invocation – The code is activated by sending a transaction to the ledger, calling its primarily assigned address and potentially transferring invocation parameters. During this step, the code is loaded from the blockchain into the executing instances, e.g. the miners.
- Operation – After invocation, the code is operated during the mining period usually in multiple distributed instances by miners who consume mining resources for operation, which usually are compensated by either the smart contract itself or the invoking instance. The result of the operation may be diverse, starting from a simple value to a new transaction up to another code deployment or invocation. After the operation, the result is added to the ledger in the consensus-process and is made publicly available.

Off-chain used in private ledgers could operate along install-instantiate-invoke.

- Install – During this step, a smart contract is made available to a peer in the blockchain system (no transaction is being written into the blockchain yet). The code itself is not active yet.
- Instantiation – The code is activated on each peer of a channel (a peer has to be selected through an address; other channel members must endorse according to an endorsement policy) and instantiation parameters are (optionally) transferred.
- Invocation – After the code has been instantiated, it can be invoked by sending a transaction proposal to one or more peers. Endorsing peers must simulate and sign the transaction and send results to the ordering service. Afterwards, the invoking party verifies whether the results are valid (checks

whether the endorsement policy has been met), batches the transactions into blocks, orders them and sends the transactions to the committing peers. The committing peers write the transactions into their local copy of the block-chain (ledger).

Essentially, both sets of procedures are similar and are analogous to standard contract law steps of invitation to treat, offer and acceptance. The legal basis of smart contracts are considered further in chapter six. However, it is the last invocation stage of the off-chain–suggested procedures that gives rise to potential concerns around the control stage. The client may retain control around verifying the transactions and vouching for the accuracy of those transactions. The challenge is how to record data in the ledger in a way that all members of the network can trust the accuracy of that data. This challenge exists even where the writing or reading of data from the ledger is handled automatically off the ledger; for example, where an "internet of things" sensor records objective measures, such as temperature or location. The ledger cannot guarantee that the automatic, off-chain sensor will always behave in the expected manner. Even in fully automated systems, power cuts, faults, connectivity outages or vandalism can break the connection between the physical and digital world.

4.8 The state of the art in smart contract development

Efforts around smart contract development seem likely to focus on the off-chain private network in the near future. The issues around scalability and privacy in relation to the public blockchain are likely to take longer to establish end user confidence notwithstanding the potentially superior benefits of the instantaneous recording and immutability of the latter.

Methods to support software developers in programming smart contracts are domain specific languages (DSLs). Foremost amongst the DSLs is the Accord Project developing open-source technology implementation for smart legal contracts. The project consists of 40+ of the world's leading law firms, along with industry bodies and corporations.

The smart contracts developed by the Accord Project[18] seek to run off-chain in a deterministic and verifiable manner, but interact with a blockchain or distributed ledger network when required. The project consists of:

1   A templating system called "Cicero,"
2   A domain-specific programming language for smart legal contracts – "Ergo" and
3   A runtime environment for executing smart contracts.

The templating system enables a user to bind natural language prose to executable logic that represents the terms and conditions of a given clause. Clauses may interact with one another and are reusable across smart contracts.

4.8.1 *Interoperability*

The key breakthrough and focus of the quickly evolving developments are around language and interoperability. Interoperability is the key outcome in terms of writing standards around the development, review and implementation of smart contracts. The current lack of interoperability between different blockchain frameworks is a major business concern and thereby promotes the off-chain versions. There are several ways currently being explored to implement interoperability such as common interchaining messaging protocols. The major challenges are around avoiding double-spending and the general trust in the link between transactions on two different chains. This may involve another evolution in professional roles such as a cross-chain manager, managing the locking and unlocking of mechanisms.

4.8.2 *Programming language*

Smart contracts are written using programming languages, where the permissible languages are limited only by the chosen distributed ledger framework(s). This option to select a programming language leaves developers free to design the smart contract and call the input and output variables whatever they want within the constraints of the language. Too much choice and too many options though can frequently be a bad thing when standardisation is required.. The greater the number of languages and codes being used, the more difficult for other to understand the code and compare it with business requirements. The International Standard records that the developer community is actively developing new languages with a little reference to decades of experience from standard organisations to deal with this issue from EDI – Electronic Data Interchange – to more recent groups collaborating on data exchange standards. ISO is responsible for a wide range of semantic standards, which would be useful in the exchange of information to and from a smart contract. For instance, a simple matter of how a date is presented in a smart contract requires there to be consensus or protocol to prevent the disambiguation of key pieces of information; for instance, should it be written 31/12/2021 or 12/31/2021 or 31 December 2021?

There are other software challenges around the development of smart contracts. A software program can in theory exist for many years. However, it is difficult to produce coding intended for a long duration where external information sources may cease to exist. The irrevocable nature of the arrangement also poses problems in terms of satisfying both parties that the coding is operating as they intended. Storage constraints, reliability and compatibility issues also require workable solutions before a wide-scale adoption of the technology is possible.

Market forces and the hype cycle curve usually see these arrangement settle down and the emergence of a dominant product or brand in any one area. The market is looking at smart contracts to function in a new and exciting fashion and this is bound to attract a degree of frenzied activity amongst start-ups out of which progress will doubtless be made. Smart contract programming will require the creation, review, approval and modification of these various approaches. It is important that business owners and developers collaborate and facilitate review for those with governance oversight. Smart contracts should proliferate and mature in the coming years.

4.8.3 Kill switches

A final question to ask in relation to the workings of a smart contract is whether or not a "kill switch" is needed. This could be presented on the basis that it is necessary to prevent any unintended malfunction or patent error by the smart contract to release funds where this was not the legitimate operation of the contract. However, the suggestion that the client has the ability to override a supposedly immutable contract could lead to supply chain nervousness that the client retains control over funds. In reality, most supply side organisation would probably bear in mind that they would be no worse off than they currently are under existing arrangements where they are effectively in the hands of the client when it comes to payment, subject to their negotiation and dispute resolution choices.

A compromise may emerge along the lines of rendering any "kill switch" decision to the internal dispute resolution provision along similar lines of a notice of objection procedure under the Dispute Avoidance Board procedure.

4.9 What might the future of smart contracts look like?

The ultimate logical extension is for whole projects to be executed from inception to completion on a public blockchain with no need for human interaction. The vision here is for contracts that are fully executable without human intervention and self-executing contract, containing electronically drafted provisions, which have the ability to automate processes in accordance with the terms of the contract.

A smart contract could be comprised of not one but thousands of mini-contracts all self-executing and transferring data as they complete and generating payment once installed and the relevant online documentation such as performance attainment and continual monitoring have taken place.

A semi-automated approach is the achievable goal in the short-to-medium term. Contracts will require judgment and the use of discretion, which requires subtlety and richness in the language which is extremely different to code. Although the benefits of full automation are diluted by this dependency on extraneous inputs, there will remain considerable value to be added by the automation of certain processes.

4.10 How might smart contracts work in the built environment?

Smart contracts, in the construction context, could achieve the automation of contract formation through to execution, updating or works programmes and the payment/certification role. The reduction in the incidence of disputes is also a major part of the business case for adoption.

Smart contracts might be the answer to the question of what can we do about the waste and inefficiencies of the construction industry. The collaborative agenda has seemingly progressed as far as it can. The huge public framework envisaged by the Framework Alliance Contract (FAC-1) is one way of achieving the goal of cutting out bad practice but this is unlikely to ever be replicated in the private sector. In the public sector, the realisation has occurred that it is only through embracing technology that the desired collaboration procedures can work. FAC-1 takes this approach in a way that New Engineering Contract (NEC) and the Joint Contracts Tribunal have yet to do.

A smart contract is a piece of executable computer code stored on a distributed ledger that, when certain conditions are met, can automatically modify data on that ledger. A smart contract potentially offers a way to enable transparent, auditable and efficient interactions between built environment businesses and government. The complex nature of the construction sector with its multitude of different players means that finding a trusted central authority or market place to approve, administrate and record the interactions will be difficult. Notwithstanding this, there is a clear application around certification and triggering work orders, which seem ideally suited to a smart contract solution.

Smart contracts could lead to huge savings in the time and resources employed in construction projects. Arbitrary deductions for retention and defects would be replaced with valuations based on the quality checks with a high degree of granularity. Issues would be flagged up and addressed at the time of construction/installation, and payment made conditional on their resolution. The whole defect liability period with the usual reluctance of the sub-contractors to re-attend site would be removed. The desire certainly exists to increase administrative efficiency, decrease the incidence of disputes and to arrive at greater security of payment for the supply chain.

Deploying distributed ledger and smart contracts within construction could see a scenario with a system wherein a supplier could trigger a smart contract designed to record the impact of poor weather conditions on the building process, thereby creating a verifiable record that a certain supplier was not responsible for the late completion of a building. That smart contract might be designed to verify the data input by the supplier by comparing it to weather readings taken from the UK Met office. The contract would make an automated call to the Met office and, depending on whether the response aligned with the data input by the supplier, would then create a record validating the supplier's claim.

Only after data had been recorded in the ledger can smart contracts be allowed to use that data to evaluate and trigger the contract. This allows smart contracts to perform their operations entirely within the copy of the database held by each node. This has the added benefit of creating an auditable record of the data being used – a record that could be used to understand the behaviour of a smart contract after the fact.

The three key challenges that face the built environment is to consider and identify how to:

i   ensure that data is inputted into the ledger in a trustworthy manner
ii deal with edge cases and resolve disputes
iii fund and govern the system, and how value is transferred

Businesses in the sector need to make decisions about the extent to which they can invest in the technology and the extent to which they can augment existing industry mechanisms to tackle these challenges. This is the challenge of designing solutions that fit the needs of the business. In the construction industry, this starts with automating performance for simple installation works inside a "stacked" contract – in part automated and in part traditionally operating.

4.11 Other possible applications for smart contracts

4.11.1 Self-drafting contracts

For a time, smart contracts had two meanings. The meaning not yet explored in this book was around contracts which were easier to use and could be put together by a lay person with inbult guidance from the contract itself. One of the drafting skills of a solicitor is in taking a basic precedent and amend if for the purposes of a particular project. The lawyer's monopoly of this role is not sacrosanct just as many other professional roles are also subject to disaggregation. This alternative meaning could merge with the wider movement towards smart contracts and offer a functionality whereby terms and conditions are selected based on machine learning as well as direction on which part of it is suitable for automation and which part not. This would result in savings in time and cost and remove the often-short termism approach to amending construction contracts to suit one parties' wishes. There could be access to an interactive library of terms, which would be used to fit the scenario considered.

The machine learning aspect of smart contract could also see the smart contract review itself for errors and inconsistencies. Discrepancies occur between main and sub-contracts due to overly complex clauses that are not always required in a "belt and braces" approach to include everything possible. This can lead to both over and under exposure of risk due to the inefficiency of manual contract drafting. The opportunity exists to rid inconsistency from the contract by taking a "click and select" approach to contract formulation. Smart contracts could offer a far more user-friendly experience allowing a far great level of clarity and understanding of the contractual terms.

The ambiguous nature of the written contract together with human interpretation through manual contract administration foster an environment of inconsistency and opportunistic adversarial action.

The inconsistencies experienced are not just between the different tiers of the supply chain. Quite often, the main contract itself has inconsistency, clashing clauses and ambiguity. There exists the possibility of running clash detection on the contract itself – to iron out ambiguities. An automated test could also check for shortcomings against legislative standards and specifications. This might have helped in the case of MTH v Eon[19] where the Judge remarked on the multiple authorship of the contract and the different agendas of those contributors. This begs the question of how often is the contract read as a whole and considered for its workability rather than each party seeking to ensure that limitations and exclusions, on the one hand, and liabilities and obligations, on the other, are safely secreted in the terms and conditions somewhere. This feature would be akin to clash detection in a BIM arrangement. Smart contracts could rely on pre-agreed logic (at an organisational level) and not an individual judgment.

Another shortcoming in current standard form contract provision is that they quickly become out of date and need to be amended. The requirement to make changes to contract documents when legislative or regulatory changes happen can be very onerous, digitisation could alleviate this. If the law changes, then it does not exonerate the user of a contract to point to their (now legally incompatible) contract and plead ignorance. Keeping abreast of legal developments is one of the key reasons that standard form contracts produce new editions. For example, JCT contracts were re-issued in 1998 to reflect the statutory regimes implemented by the Housing Grants Construction and Regeneration Act. A smart contract could ensure its own legal compliance by updating its terms against a central network of regulations and precedent.

4.11.2 More accurate planning and delivery

Machine learning offers the ability to check against other scenarios and completed projects to extricate the relevant predictive data for a new project. The processing power available with the latest computers is phenomenal and provides the ability to run simulation modes of any obstacles experienced. The greater granularity afforded by new shared data sets potentially *enables* the agreed reduction of priced risk contingencies. The essence of a standard form should be to integrate into all aspects of project delivery by embracing technology, facilitating the execution of transactions and reducing disputes. The potential is there to add efficiency and clarity to the construction process. The contract should no longer be the "rules of engagement" by which battles are fought but a proactive guide to efficient and successful project delivery.

Already programmes such as Contract Event Management and Reporting (CEMAR)[20] offer much better informed contract management and auditable records. In a similar way to how dispute avoidance recognises that prevention is better than cure, just so that a smart contract can seek to optimise its own project performance and how it will perform in the future. The programme could be regularly running in simulation mode to test situations. This optimisation approach is already being seen in other sectors such as energy supply. A UK-based company called Electron wants to use smart contracts on the Ethereum Blockchain to develop a smart grid that will always deliver energy. Energy can automatically be routed via the optimal way through the grid for the most efficient grid use and least energy loss. This data should be public as every grid node could have its own (financial) interest in getting most of the energy through its own part of the grid. The technology is currently in the testing phase with 53 million metering points and data of 60 energy suppliers.[21]

A similar application in construction is conceivable where the smart contract can make more transparent the activities within a supply chain by tracking each step and each transaction between multiple contracted and sub-contracted parties via the internet of things and the submission of electronic information by parties to the smart contract and/or agreed upon oracles. This can ensure optimised supply chains and inventory tracking and timely delivery as well as enhanced tracing and verification in order to reduce the risk of fraud and theft and allow for timely insurance payments when these do occur.

4.11.3 Addressing climate change

It is possible to see an application for smart contracts where they specifically address the societal value agenda. The volatility of crypto-currencies means it is unlikely that sufficient confidence will be generated in these tokens in the short term. However, other token approaches would amount to an ideal test bed for these technologies. A green token based on auditable big data records of the provenance of a component to a building project is perfectly feasible. This token approach can continue through the construction phase of the component's life and in-use data can be recorded alongside the component's delivery cycle. This resonates with the circular economy,[22] which aims to eliminate waste and re-use resources. This regenerative approach is in contrast to our traditional linear economy with its "take, make, dispose" model of production. The role of the designers in bringing this about is essential. They can supported in this by auditing and provenance checking through blockchain and smart contract technology.

The work of the Gold Standard[23] already has created a system of monetarising "carbon credits" and it would seem like a natural progression for such organisations to seek to further automate their procedures through blockchain and smart contracts. Smart contracts also have a role to play in meeting the United Nations Sustainable Development Goals. Smart contracts indirectly contribute to all of the goals. The most closely linked are goal 9 – industry innovation and infrastructure, and goal 11 – sustainable cities and communities. Data is the key and smart contracts act upon that data. For example, a company called Provenance provided a case study around tracking tuna fish and their journey from shore to plate. The benefits of this approach was to ensure sustainably sourced fish and a fair deal for those involved in the road to market. *"All the kinds of problems that certification and audit aim to address could potentially be remedied or streamlined a huge amount by blockchains and smart contracts, and I think that is why everyone is getting very excited about it."*[24]

4.12 Why have smart contracts not yet arrived?

Smart contracts are currently being hyped and are therefore currently on the way to the peak of their inflated expectations buoyed by claims as to the benefits. There are plenty of potential pinpricks, which could see the bubble burst. The public blockchain suffers from reports about its limited scalability, low performance and lack of privacy. A survey into the main barriers of blockchain adoption highlighted regulatory uncertainty and lack of trust among users as the two greatest reasons for non-adoption. The survey also predicted the likely obstacles in 3–5 years as being the questions of cost, how to start and lack of governance.[25]

4.12.1 *Late adopters*

The takeover of the world of mainstream transactions by smart contracts has been predicted for several years. Even in the most likely of sectors, financial services, there developments are only piecemeal and the breakthrough to reality is limited. This is the province of the start-up company that can boom and/or bust very quickly. This is particularly true for those businesses tied to the volatility of a crypto-currency.

Financial service agreements are much more suitable for smart contract transactions than construction projects. The former consist of short-term transactional-based exchanges of instantaneous effect where one thing (money) for another (asset). The adoption of smart contracts in this area is known as FinTech. However, this is at odds with the complicated and long running nature of construction agreements. This gives credence to the contention that the stack approach taken by Allen is likely to prevail in "nesting" automated provision inside more traditional arrangements. Further, storage constraints, compatibility and reliability issues together with confidentiality and the long-term nature of distributed ledgers pose additional problems.

The take up of technology in the construction industry languishes behind other industries. The sector is currently making heavy work of adopting the BIM, which is the gateway to digital design and construction. The learning curve is quite steep and requires investment to achieve BIM familiarisation. Smart contracts present a complimentary technology with the bonus of being easier to grasp and utilise for stakeholders. Certainly, the technology exists but not the confidence in these arrangements. Confidence stems from there being a proper legal foundation for their adoption and operation.

4.12.2 *Adequate software development*

Smart contracts are seen to come from the same stable of measures as the blockchain and crypto-currency markets. There are also viewed with the same suspicion in terms of legal uncertainty. However, this is unfair as the smart contract strictly depends on neither. The way in which smart contacts are delivered is not settled. There is no reason why the public blockchain should be used for what is essentially a private matter between the stakeholder actors. A private ledger approach appears much more likely.

Smart contract platforms suffer from serious issues such as bugs and security holes and they are not scalable enough to support business applications. New technology is rarely perfect and things will improve over time. The idea that the code written will be flawless and safe from exploitation is also questionable without the benefit of prototype testing. A prominent example of the risks around immutable coding was demonstrated around an automated investment vehicle known as the DAO – Decentralised Autonomous Organisation.[26] Investors pooled their money in the organisation and were able to deposit and withdraw funds as they saw fit. A flaw in the system meant that there were no checks whether an individual had withdrawn more than their original investment. A certain user or users exploited this function to remove $50 million from the fund by hiving it off to a separate fund. The community of investors cried foul notwithstanding the hacker's claim that they had merely followed the allowed procedures of the DAO and were within their rights to keep the funds. Ultimately, the matter was resolved by short-circuiting the supposedly immutable code and its accompanying consensus mechanism. This is known as a "hard fork" in the blockchain, which serves to wind the clock back and create a new chain of transactions. Confidence and share price in such automated investment platforms plummeted as a result.

And yet, by taking a longer-term view of such setbacks, the processes will become more robust as a result. The common law is the wealth of experience and the incremental development of a set of rules to govern all the situations that have come before the court. The DAO hack actually performed a service to the development of the ledger technology and its procedures and safeguards, which were no doubt updated and made more secure as a result.

Even where smart contracts execute based on accurate data, there will be times where they do not function as expected. These are called "edge cases." Edge cases can arise from changing real-world conditions that invalidate their function, such as supplier in the supply chain going out of business. Also reports of malicious or unintended exploitation of flaws in smart contract code.

One often-cited reason for these vulnerabilities is that fact that the languages and environments that smart contracts are written and operate in are relatively new, and are in many cases still under development. This is why the current task occupying the pioneers is the software development. Once the most functional language has been written and receives the appropriate backing, these issues should decrease over time.

Representing the real world though is likely to remain an issue for smart contracts for some time. This could hasten the adoption of factory built and prefabricated components created in a replicable and stable environment that is not at the mercy of site constraints and extended supply chain permeations. This should lead to automating business processes ultimately making them less vulnerable to exploitation.

Another factor that contributes to the vulnerability of smart contracts is that they are difficult to test before deployment, especially when they interact with other contracts or real-world processes. The dearth of any sufficient research and development money in construction means that the testing and prototyping that should be being carried out awaits the appropriate investment. In other industries, it would be inconceivable not to invest heavily in research and development to investigate the desired outcomes.

4.12.3 Representing the real world

Data has already been identified as the essential ingredient of smart contracts. This means that each application of a distributed ledger needs some form of data to be recorded into the database. The recording of the data into the ledger is required to create an immutable record of some real-world event or interaction. The fundamental question is raised of how the distributed ledger can represent data about a thing or event in the real world in a way that ensures the data being stored in that distributed ledger is an accurate representation of that real-world thing or event. The mechanism for injecting data into the ledger in a trusted manner involves the intermediary of oracles and data analytics.

Another limitation here is that we are a long way off being in a position where the smart contract itself can access off-chain data sources directly. The nature of the distributed ledger is that all the nodes involved would have to be updated to reflect the smart contract and the fulfillment of its contractual terms. If this were a public ledger, then that would mean all 17,000 nodes on a public Ethereum network being updated for every event. The only way around this is for the data about real-world events to be in place before any smart contracts attempted to execute their terms. The smart contract must therefore be able to predict and have contingencies for all possible outcomes. The alternative is to have an exception-handling construct such as the oracle.

One way that the real world is represented in contracts is through the elastic concepts such as reasonableness. This can be deemed intentional ambiguity in contracts. Software writers think in terms of exception handling and want to address bugs at compile time rather than run time. Finding a bug in a contract at run time is called litigation and is very expensive.

The word reasonable can be viewed as being a function call from the domain of the automated contract into the domain of humans. A judgment call on whether or not something or some action was reasonable could be assigned to a "semantic" oracle. The human would need to check the traces of the action and answer the question was that reasonable – yes or no. The human does not need to be a judge it could be the expert witness or someone both parties have agreed ahead of time would be the decider.

4.13 Conclusion

The working position arrived at is that certain aspects of the construction contract cannot be fully smart and the best that can be achieved in the short to medium term is a semi-automated position. Further, smart contracts should be viewed as part of the BIM-led revolution in construction and not separate from it. The recommendation is that incremental advances such as the coding of project management and contract administration data be targeted to provide improved operational efficiency and value savings.

Trust and money were the key words for Latham. Smart contracts have a claim to provide both in sufficient quantities to make a real difference to the construction industry and end decades of inefficiency, waste and disputes. Chapter five examines the views and perceptions of construction industry stakeholders to these developments. The buy in of those who would be the users of the smart contract movement are clearly important to consider.

Notes

1. Laidler, K. (1998) *To Light Such a Candle: Chapters in the History of Science and Technology*. Oxford University Press, Oxford, UK.
2. Kurzwell, R. (2014) *How to Create a Mind*. Duckworth Publishing, London.
3. Raskin, M. (2017) *The Law and Legality of Smart Contracts* I Geo, Law and Technical Review 304.
4. Savelyev, A. (2016) *Contract Law 2.0: "Smart" Contracts as the Beginning of the End of Classic Contract Law*, Higher School of Economics Research Paper 71/Law/2016.
5. Allen, J.G. (2018) *Wrapped and Stacked: "Smart Contracts" and the Interaction of Natural and Formal Language*, European Review of Contract Law 14(4), 307 at 313.
6. Szabo, N. (1994) *Smart Contracts*, Unpublished Manuscript.
7. AA v Persons Unknown who demanded Bitcoin on 10th and 11th October 2019 and others [2019] EWHC 3556 (Comm).
8. Nakomoto, S. (2008) *Bitcoin: A Peer-to-Peer Electronic Cash System*.
9. Blockchain and Distributed Ledger Technologies ISO/TR 23455:2019 available from www.iso.org.
10. Institute of Civil Engineers (2018) *Blockchain Technology in the Construction Industry*, available at: https://www.ice.org.uk/ICEDevelopmentWebPortal/media/Documents/News/Blog/Blockchain-technology-in-Construction-2018-12-17.pdf last accessed 09/07/2020.
11. https://www.iso.org/iso/iso_strategic_plan_2011-2015.pdf.
12. Wood, G (2018) *Ethereum: A Secure Decentralised Generalised Transaction Ledger*, available at: https://gavwood.com/paper.pdf.
13. Lessing, L. (1999) *Code: And Other Laws of Cyberspace*. Basic Books, USA.
14. Norton, Rose & Fullbright, (2016) *Coding the Fine Print*, available at: http://www.nortonrosefulbright.com/files/smart-contracts-137872.pdf last accessed 07/07/2020.
15. https://www.ashurst.com/en/news-and-insights/legal-updates/smart-contracts—can-code-ever-be-law/.
16. R v Craig and Bentley (1952) *The Times*, 10 December.
17. This does not work on public blockchains which do not support methods to change smart contract behaviour. In a very extreme case, it is possible to fork a blockchain transaction history, giving the community the option to adopt any of those forks. This was famously used in the Ethereum blockchain following a hacking incident. See Blackburn, C. (2020) *Ethereum Blockchain Revolution Explained*, American Bar Association.
18. The Accord Project, www.accordproject.org.
19. MT Højgaard A/S -v- E.ON Climate & Renewables UK Robin Rigg East Limited [2017] UKSC 59.
20. Cemar.co.uk.
21. Electron.org.uk.
22. www.wienerberger.co.uk, available at: https://www.wienerberger.co.uk/tips-and-advice/sustainable-building/a-circular-economy-for-the-built-environment.html.
23. Gold standard, www.goldstandard.org.
24. Jessi Baker, Founder of Provenance.org, Case study from shore to plate: Tracking tuna on the blockchain.
25. Institute of Civil Engineers (2018) *Blockchain Technology in the Construction Industry*, available at: https://www.ice.org.uk/ICEDevelopmentWebPortal/media/Documents/News/Blog/Blockchain-technology-in-Construction-2018-12-17.pdf last accessed 09/07/2020.
26. David Siegal, *Understanding the DAO Hack* (coindesk 25 June 2016), available at: www.coindesk.com/understanding-dao-hack-journalists.

5 Perceptions in the construction industry of smart contracts

5.1 Socio-legal studies

This chapter draws together three studies examining the views and perceptions of construction industry professionals. Taking the temperature of current attitudes towards smart contracts in this manner is an important step and prevents the discussions from becoming too disconnected from the eventual users. In the first paper, secondary data was collected from online fora where the identities of the contributors are unknown to the viewer.[1] Three prominent contributors to the debate were followed and their responses were studied.[2] Contributors A and B were chosen due to their sceptical attitude towards smart contracts. Contributor C had a much more positive attitude and appeared to work with blockchain technology. In paper two, the author and a quantity-surveying alumnus, Hollie Escott, wrote about the views and perceptions of over one hundred stakeholders by subjecting the findings to thematic analysis.[3] In the third paper, current PhD candidate and former International Construction Law alumnus, Alan McNamara, conducted a study with his supervisor to ascertain views on the central question of the benefits that smart contracts may deliver.[4] The findings of all three papers are presented below. This data is not presented as being necessarily conclusive or as an evidence for any particular prediction made. The aim is to portray a sense of the underlying attitudes and the range of understanding towards smart contracts and information technology more generally. The third study was conducted in Australia, which has created a reputation for itself around technological advancement and innovation, particularly in the built environment fields with a good deal of cutting-edge research activity.

All three papers seek to contribute to the socio-legal approach where law is realised at the point at which it interfaces with its users. Positive indicators amongst a predominantly reactionary industry are particularly noteworthy and there are encouraging signs where people are willing to embrace both optimism and innovation for the industry. The terms are defined thus:

- *Optimism* – The belief that technology offers people increased control, flexibility and efficiency. The belief that a reduction in human error can be achieved with the automation of tasks and the provision of increased quality assurance through certification and verification of coding through digital ledgers feeding into a smart contract. The belief that smart contract would act as a trustworthy contract administrator by introducing an error-free process based on the contracts would be both built and monitored.

- *Innovation* – The appetite at an individual or organisational level to pioneer and provide thought leadership across an industry. The conservative approach taken by many in the industry could be seen as an opportunity for other more forward-thinking companies to innovate as well as for players outside of the construction industry to seize opportunities.

5.2 Qualitative responses

Observations on smart contracts are made in this section interspaced with indicative views from construction professionals obtained from a secondary data source. Smart contracts are deemed desirable because they will save cost and time in automating certain aspects of construction project performance while saving cost and time in transaction arrangements. The potential to provide ongoing as-built information for use in whole life costing is also attractive for clients.

Recognising the positive effects achievable through smart contracts is not universal. Contributor A to the online forum stated: *"decentralisation [seeks]... to replace people as they view as unnecessary middle-men. The absurd autonomous organisation idea is an extension of that concept."* The same contributor continues: *"Smart contracts are the stupidest misunderstanding of the world the blockchain movement has produced yet."*

Contributor B was not so dismissive of the central notion but succumbed to the temptation with contemplating future technology is to take things to their logical conclusion without question. *"I guess maybe someday we'll have a god like AI.... but we're not close and I'm not convinced that it's possible."* Contributor C was more positive and made the comment *"complexity is not insurmountable."*

The discussion on the forum addressed the point about whether intelligent contracts can deal with the uncertain elements in a construction project and resolve the ensuing disputes. Contributor A: *"Disputes do not revolve around inaccurately recorded information and it is not the case that having a secure ledger to record transaction/contract details would solve most problems."* Contributor B agreed by making the point: *"genuine disagreements between two reasonable people can't be solved objectively by a computer."*

Respondent C had a different approach stating:

> of course parties can be intractable and a course of action dealing with that condition would be required, but if agreeing upon dispute resolution procedures and third-party consultants as part of the construction contracts is common practice now, why wouldn't similar measures for reducing disagreement be applied in automated contracts? We are teaching computers how to paint, why can't we teach them how to lawyer?

The discussion returned to the semi-autonomous advancements possible through smart contracts. Contributor A and B recognised that record keeping and payments could be regulated through smart contracts although remained dismissive of these benefits: Contributor A: *"If all the smart contract can do is move money around when clearly objectively defined conditions are met, then it's just an automated escrow; big deal."* Contributor B: *"Ultimately all a computer can do is check some numbers against some other numbers. Human interaction is generally more complicated and nuanced that that."* Again, it fell to Contributor C to recognise the potential here: *"When those numbers are certified to represent identity, authority and agency it is a rather powerful combination."* On the disputes point: *"Avoiding lawsuits and reducing instances of substantial disagreement is an excellent use for immutable distributed record keeping… When facts are harder to dispute, there are less disputes."*

Looking further ahead, Contributor A remained as the voice of portentous things to come: *"You ever see Dr. Strangelove? The Soviets had a 'smart contract' that said 'if nuked, destroy world' turned out to be a bad idea."* Contributor A continued: *"There is also one major threat with every custom smart contract: if there's a bug, you're permanently screwed."* Contributor B took a more pragmatic approach to the shortcomings: *"Please tell me where the information a 'smart contract' uses to process comes from. Answer: from some arbitrary APIs (Application Program Interface) which may not be accurate, or may not be updated at the right term, or any other thousands of issues – malicious intent, spilt coffee. There's an old software adage that goes 'garbage in, garbage out' that is it down's matter how beautiful or well written your code is, if you receive garbage input, you'll invariably have garbage output."*

Contributor C who re-enforces the potential for smart contracts:

> Blockchains can increase the authority, security and transparency of record keeping systems and smart contracts can enable interoperability with other systems such as payroll, insurance and supply chains. A major cost driver of public projects is accountability oversight. If specifications, bids, Requests for Information, purchase orders and delivery schedules are automated, validated and transparent on a globally accessible blockchain… then tracking costs as well as the redundant overhead can be reduced and theoretically corruption would have less places to hide.

Contributor C focuses instead on the incremental improvements that smart contracts can make in terms of project savings. *"A smart contract for escrow can eliminate the need for a third party as a temporary possessor of funds or deeds, reducing risk exposure, service fees. It can accelerate processes from days, weeks and months, down to seconds, minutes and weeks depending on the complexity of preconditions for closing."*

5.3 Quantitative responses

Almost half consider their company to be innovative and at the forefront of new ideas, a striking statistic considering the industry has a reputation for being slow to change and behind innovation (Figures 5.1–5.3). The next question asked is about the knowledge of smart contracts (Figure 5.4).

Figure 5.1 Respondents by profession

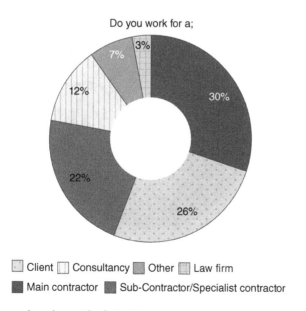

Figure 5.2 Respondents by supply chain position

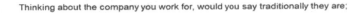

Thinking about the company you work for, would you say traditionally they are;

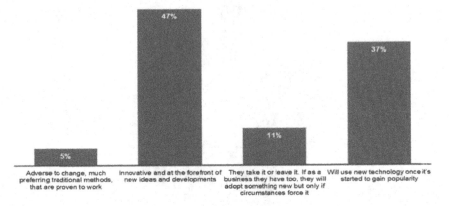

Figure 5.3 Typical attitudes in the construction industry

The knowledge gap identified will necessitate a lengthy and costly education to ensure people fully understand the technology and its potential before considering adoption. Section 5.4 examines the respondents' comments, which accompanied their selections on the questionnaire. This is accompanied by the author's commentary based on the literature reviewed.

Prior to this survey, had you heard of and how much did you know about Smart Contracts?

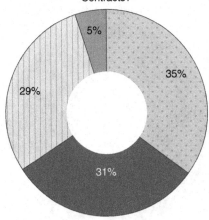

☐ I had never heard of them until today

☐ I know a little about them

■ I've heard the term, but not really sure what they are or how they work

■ Yes, I know all about them and how they are being developed in other industries

Figure 5.4 Knowledge of smart contracts

In all, the respondents made 300 comments to accompany their multiple-choice selections. This discussion features a selection of these insights, which have been grouped together in a thematic analysis. The purpose of setting out the quotes is to allow the reader to select and reflect on those sentiments, which resonate with and potentially challenge their own stance on technology.

The study is akin to taking the temperature of a patient during a health check. The responses captured a reflection of the industry.

Taken as a whole, the construction industry could be characterised as being a cynical lot. Professor Mosey addresses these comments at the cynical:

> A healthy amount of cynicism can help to avoid wasted time or wishful thinking, an excessive dose can be toxic. Do not undermine proposals because you assume they are not new, or are naïve, or are too risky or disguise someone's hidden agenda.[5]

5.4 Responses generally

The importance of definitions in legal scholarship and contract writing is of great importance. This discussion is no exception. The respondents apply succinct and largely accurate responses to the question as to what they understand a smart contract entails. One could surmise that, given the richness of the definitions and the paucity of knowledge claimed, the respondents turned to technology (internet searching) to augment their own definitions.

> A contract performed by technological means. Once initiated, a smart contract is typically irrevocable and automatic. It can either be a coded contract or compromise a written contract which has automated functionality – such as payment- in smart contract alongside a more traditional written contract.
>
> I understand the concept of instantaneous payment and a "brick by brick" idea of plotting progress. Also the idea that the middle men are removed from the process such as banks.
>
> The general premise is that they are pre-programmed and have the ability to self-execute.
>
> They are aimed to facilitate and project manage a given contract on behalf of the user with his or her minimal input and supervision.
>
> They are computer programs that take the place of traditional contracts for simple contractual arrangements primarily used for digital transactions.

The remarkable point is how accurate and seemingly well informed the respondents are on a subject about which the majority profess not to know anything. The ease with which the subject can be grasped therefore, sometimes by deduction alone based on its logical premise, bodes well for long-term adoption.

The comments start with a recognition that it is incumbent on the client side to make the investments of time, money and energy into innovation.

> Smart contracts will only truly succeed when people and organisations change their procurement mind-sets and strategies and invest more in innovation.
>
> As a client organisation we should lead the industry into the digital age and insist these items are taken up now, even if just in small ways.

The comments selected move on to point out the shortcomings of the status quo in terms of contract drafting.

> Contract drafting remains very clunky and expensive and clients dislike it, and yet they derive comfort from the to and fro of negotiation.
>
> There is still far too little standardisation with contracts in some industries. In my industry I have not received any contracts based on standard forms in the last two years.
>
> Where standard forms are used they are often modified to the point where they are no longer standard leading to potential for ambiguity and disputes. If smart contracts is used correctly it can only be beneficial for the construction and manufacturing industries.

These comments sum up the shortcomings of the current legal arrangements whereby a great deal of resource is used in the negotiation of contracts and seemingly self-defeating nature of amendments. Any steps towards the automation of the process would reduce this expenditure.

5.5 The positive comments

5.5.1 *The potential for positive impact*

There are 81 recorded comments in this section featuring repeated themes and some similarities in approach and expression. Not all of the comments are set out in this discussion. The author has sought to capture the essence of the insights shared by the respondents. Nine themes have been used for the purposes of analysis, they are:

> The key sentiment being expressed in this theme is that it is not necessary to know how the smart contract works in order to use it and trust it.
>
> A car can be driven with great skill without knowing how the engine works.
>
> Everyday users of smart contracts will not be exposed to the coding structure and algorithms needed for it to work.
>
> If proven to work, the code is likely to be less of an issue.

An observation here is that the time has now passed when a construction professional needed to know the intricacies of the process and their involvement in it. This loss of specialist knowledge is, on the one hand, regrettable for those of us brought up on the discipline-based approach. On the other hand, younger generations would see this as entirely normal.

5.5.2 What to do about the bugs

The respondents' experiences around information technology and its reliability are addressed in this section. The prevailing view is that bugs can be worked around and should not prevent progress per se.

> All bugs are manageable through the transition phase.
> Anticipating bugs is a very backwards opinion and could be said of any technological advances we've made.
> As technology resilience gets better there will be less bugs/issues. Bugs will not necessarily cause disputes.

The last point is particularly reassuring for those seeking to promote smart contracts. Building Information Modelling has its detractors, based on its complexity and the need for clients to be clear on what they want. The investments made by the technology providers has provided a degree of reliability. This investment is rewarded through the licence fees charged.

5.5.3 The relationship between smart contracts and trust

The respondents were questioned on whether trust and collaboration were "baked in" to the smart contract process. This section analyses those responses appreciating that a transparent and granular process can actually add trust.

> The building of trust and collaboration is a key benefit. This is good for building more long-term strategic alliances with your suppliers – more cost efficient savings to be made.
> If the technology really works and people give it a chance then faith and trust can be built. If things were more automated and out in the open and clear for everyone to see then maybe the trust between the parties would improve.

These comments appreciate the point that the transparency of the procedure would result in the trust required and the benefits presented by a long-term strategic alliance. The last comment is insightful in demonstrating that the trust is already in place well before the smart contract is required to manage the transactions.

5.5.4 The benefits of standardisation

These comments recognise the benefits of standardisation in terms of ensuring minimum quality standards and the potential to reduce disputes. The removal of subjective wording is also seen as being beneficial in this section:

> The potential to remove errors in drafting contracts and avoid individuals interpreting contract clauses differently due to the way they are written could be a major advantage of smart contracts.
> The technology can help resolve disputes on timings and what has been agreed, issued, paid etc.

The difficulty is that the replacement of subjective wording, although desirable, with computer code is a considerable ask in terms of being able to cover the variables encountered on a building project. The point is made that the reason for having a written contract in the first place is to deal with the uncertainty arising from the project (see discussion section 5.2). This limitation on what is achievable may be partially overcome by an uptake in pre-fabricated or factory-type installations on individual or linked elements.

5.5.5 The business case for adoption

The notion of smart contracts requires the business case to be made out. The respondents in this section were quick to recognise the positive impact this technology could have.

> As a subcontractor we can spend a lot of time chasing payments so a system to reduce this work load would be of benefit to us.
>
> As with innovations like BIM the company are happy to invest in order to be market leaders in new technology.
>
> Cost savings on administration may be moved over to IT support to solve a unique bug to a specific contract or a system bug that assists all of the companies.
>
> If the Managing Director of a company believes in the idea then a top/down attitude will spread across the business and give a new idea a very good chance of succeeding.
>
> Less paperwork and more automation will eventually reduce writing, negotiating, operation and admin costs. This simplifies the process of stage payments and administration of this process from a contractor's point of view.
>
> Subcontractor payments are one of the industry's biggest problems anything to improve payments has to be a good thing.

A recurring theme here is the potential for redeployment of people and resource to other tasks. The huge amount of time and effort spent in chasing payments appears as a frustration to the respondents. The references to "cutting out middle men" are of interest. Most seem not to appreciate that the middlemen being cut out in this instance may be their colleagues and peers.

The last comment identifies one of the major benefits of the smart contract initiative. The poor treatment of subcontracts and the weekly cost of this in terms of insolvencies are stains on the image of the construction industry. The alleviation of this through a transparent and repeatable payment process is to be welcomed.

5.5.6 The relationship with BIM

The literature reviewed established that smart contracts probably need to reside alongside BIM, as this is where the information will be harvested in terms of the specifications and standards to be achieved and verified before making the payment. The equivocal nature of this prediction is based upon the observation that "anything goes" in relation to smart contract development and what it will be tethered to both now and in the future.

If BIM is the pathway chosen then there are encouraging signs as remarked upon above that the experiences are positive.

> BIM is also growing and an exciting way to standardise construction data.
> Improvements in technology including BIM will deal with issues before they arrive on site saving cost and time.

The observation here is that in a survey of 117 results, only three commented on the BIM position, which speaks volumes on the relatively small numbers with BIM experience.

5.5.7 *The process of automation*

Automation can be distinguished from standardisation, as the latter is one of the benefits from the former. The process of automation itself was addressed in a positive manner by the following respondents:

> Clearly emotional issues cannot be resolved but the objective decision as to the right process after an event can be solved by a computer.
> Disputes that are likely to occur will still centre around risk or scope and there are likely to always be work arounds that can be deployed to make the process work.
> Technology brings greater accuracy from survey devices to drawings and cross checks/approvals.

The last reference to the application to manufacturing processes echoes the constraint about the vast scope of variables that would be experienced on a wider construction project.

5.5.8 *Other positive comments*

This section identifies some of the other positive comments made before the discussion turns away to look at some of the more intractable issues. The comments underline the "buzz" around the subject, which supports its position on the "Hype Cycle".

> This is exciting as it brings new talent and ideas, innovative as it changes age-old practices better because it could simplify contracting build trust and reduce disputes. What's not to like?
> Never underestimate the ability for people to adapt their methods of working and communicating (social media did not exist 20 years ago).

These are exciting and innovative and could change the way we work for the better.

There are many improvements that can be effected in the construction industry by technology to give managers more time to deal with "real" issues that need thinking time to resolve.

Wait and see could lead us to stagnate whilst waiting for things to improve. We can't be afraid of trying and failing and trying again.

There are a good deal of positive statements here and therefore plenty of support for the process. Presumably, those advocating the redeployment of the labour away from the tasks that would become automated are mainly at a managerial level, where their own role is not so compromised. The insight demonstrated in the last two comments underlines the potential to revert to existing methods in the event of issues arising.

The position reached with the sum total of the comments examined so far is the highpoint of the hype cycle. The business case, potential and technological readiness have all been made out together with a good dose of respondent enthusiasm for the process. These positives should be borne in mind for what they represent – a compelling case for smart contract adoption. Section 5.6 of comments examine the limitations present on the technology, which has the effect of dampening these expectations.

5.6 Not-so-positive comments

5.6.1 Control

These initial comments in this section reveal the perception that the controlling and opportunistic behaviour in the construction industry will remain of pivotal importance despite the advent of automation.

I don't believe it will ever be fully automated managers will always like to keep control of payments and not rely on the payments being automatic.

A contractor will always want to maximise profits and therefore seek CEs.

In larger value disputes one of other party will want to take things as far as they possibly can and are unlikely to accept a computer's verdict if they are the party found against.

The last two comments are similar and have the ring of truth about it in terms of parties wishing to have the control to escalate a dispute which can occur sometimes regardless of the merits or advice received. A lack of control or a surrender to an automated third party is identified as a limitation here in terms of the ability to act commercially. Expecting institutions to act against their vested interests is fanciful. The business case must therefore be made out to demonstrate a new way for those interests to be protected.

5.6.2 Lack of a complete solution

The potential is recognised for improvement here. The smart contract initiative is effectively damned with faint praise about its limited potential impact.

> A reduction in paper is always welcome but the reality is that storage of a document would still be required and need to be retrievable.
> In a busy office would you revert to the "old" systems – new technology will only be used if easy and user friendly.
> Ultimately due to the fact that this will not eliminate human requirements in the process the most beneficial aspects would be reduction in errors and speed.
> You will still need a QS to do a role, it's just QS could add more value now, as opposed to paper pushing.

The process, when viewed collectively through these comments, still appears to be extremely worthwhile even if these limitations are accurate.

5.6.3 Too many variables/complications

The main drawback of the process is featured here in the limitation section and returned to in section 5.6.4 where the sentiment is expressed not only as a limitation but also as an insurmountable obstacle.

> All the little bits of variation from the millions of random events have to go somewhere like scope or budget. You can automate these events, but only if they are repeatable; these characteristics of logic and maths leave little space for smart contracts.
> Breaking down the components of construction cost (Resources employed) and the item/element of work to its lowest common denominator will be challenging. There will inevitably mean higher risk built in and consequently a higher charge to the client.
> In my opinion full automation would be relatively simple to achieve, however I have concerns over how you would automate dispute resolution and issues that arise once work has started on site.

The next limitation is linked to these observations, which examine the inability to scale up the process from simple manufactured based tasks.

5.6.4 Simple tasks only

> Full smart contracts would suit factory type production or linear projects or those with precise and clear activities.

The last comment holds out some hope for the process in terms of rather than seeing it as being only suitable for simple projects the solution could lie in viewing the whole as a sum of its parts. The individual process can therefore be broken down into its constituent processes.

5.6.5 Human contact required

Again, these concerns are expressed all the louder in section 5.6.6. A semi-automated process is being anticipated in this latest limitation.

> A lot of works in the construction industry is undertaken on trust and collaboration break the trust and the project becomes a contractual mess. Would a computer system have trust?

The last view raises the point that the role of the human is to enthuse. There may well be something in this point although reducing the human to a "cheer leading" role is to become marginalised in importance.

5.6.6 Lack of a driving force

This final limitation identifies the requirement for intervention in an industry well known for its resistance to change. The parallels are there to be seen with BIM and its limited take-up in the private sector despite it having been mandated in the public sector. This section has analysed the responses concerned about the limitations of smart contracts. The main ones address the complexity and requirement for human intervention in the process. The clamour of the naysayers builds to a crescendo in section 5.6.7 where the scepticism around the claims of smart contracts is addressed.

5.6.7 Cloud-cuckoo land

This section is reserved for the out-and-out doubters. The points raised are nonetheless revealing and contain valuable insights. The accusation of doublespeak can, in part, be upheld. A contract cannot be smart in the sense of real intelligence. The misnomer seems to have stuck, however, in the way that BIM does not really cover the process it has become. The remarks about deteriorating relations between main and subcontractors are also noteworthy against a background of Brexit nervousness in the economy. Further, the accusation of being too far removed from site level provides a caution against being divorced from the realities of the construction site.

5.6.8 IT fatigue

This section examines the comments based on people's ennui and mistrust of computers usually borne out of their own bad experiences.

> Anything that is IT developed is prone to hacking and similar system errors. Technology is fallible as are humans. Technology which is poorly applied may well cause more initial disputes but because it is linear disputes should be resolved very quickly if they have arisen due to technological errors and bugs etc.

> The risk of coding incorrectly is a big risk throughout the process from implementing to completion. This would inevitably lead to change.

Some of the comments made above are phrased as questions which, once answered satisfactorily, would address the concern. The back-up of the system would be a reliability issue addressed in the development stages of the technology. The "rubbish in, rubbish out" phenomenon has not tainted people's experience of BIM and those standards would also be achievable on this platform. The comment about being able to unpick linear disputes arising from computer error is actually a positive. This challenges the received wisdom that "to err is human, to really mess up requires a computer." Quite the opposite would appear to be the case.

The final comment is reflective of one criticism levelled at IT suppliers – namely, the need to constantly upgrade the systems and pay additional premiums. Smart contracts would not appear to be in a position to reverse this trend.

5.7 Indifference towards smart contracts

The old saying about it is easier to tear down than to build up should preface these remarks.

5.7.1 *Claims on costs unfounded*

The first challenge to the smart contract project is that the cost savings case is not made out.

> I'm not sure there would be cost savings on a typical contract as the backlog of unresolved issues would lead to more cost to resolve.
> The amount of cost savings I think will be unjustified, as system install and upskilling will null that out.

In the main, these comments are not denying that cost savings are possible, merely that the case has not yet been made out sufficiently well. This is something that can be addressed either academically or commercially.

5.7.2 *Contractual purpose*

These scepticisms centre on the unique ability of written contracts to be able to cope with uncertainty generated on construction projects and to be a frame of reference for their resolution.

> Construction contracts cannot be 100 percent prescriptive the uniqueness of the construction delivery cannot be "beyond all reasonable doubt."
> If the problem is complex enough to require a contract, it's too complicated to predict the outcome.

The point being missed here is that the nature and content of a contract will be vastly different in the envisaged project. The contract will be prescriptive and exhaustive in its handling of the variable scenarios to which it will apply. This will be achieved either at the component or trade skill level. This leads into the next criticism, which is whether that level of detail is ever going to be achievable.

5.7.3 Too complex/unique

This section revisits the complexity issue already identified as a limitation. Here, the limitation becomes a major obstacle.

> Construction is a very complex business where there are many risks and changed circumstances which need to be well managed to compute with a variable option for each event encountered based on yes or no answer I cannot see happening.
>
> It is very difficult to produce an automated response no matter how cleverly coded to cope with all anomalies that might/can occur.
>
> Too many "what ifs" for coding to ever reach a point of agreement. Computers now far more capable but the concept is still flawed because of the immense variable inputs – let alone the output expectations.
>
> Even with the immense progress that has been and will be made in technology the complexities are not to be underestimated and human attitude is what is holding things back more than anything.

A recurring theme in the above comments is that construction projects are unique. Clearly, the size and complexity of projects vary but the degree to which they are truly unique is open to challenge. A more open debate on the topic can be achieved by approaching from a different angle – all construction projects involve elements of repeatable and severable processes that can be automated. There are a glimmer of hope in the last comment which identifies it is human attitude about what computers can do which is the obstacle.

5.7.4 Risk of replacing humans

This scepticism follows on from the points made about the complexity of the process. The perception is that humans can deal with the complexity and prioritise what is important. One of the most important things is held out as trust and collaboration which cannot, so the thought goes, be replaced by computers.

> How can code replace human judgment interpretation and variable management techniques? Removing the human element will reduce trust and collaboration.
>
> You need interaction for trust and collaboration and this does take some of that out of the process.

The last comment acknowledges the sea change necessitated by the smart contract project. The construction industry has endured twenty years of the promotion of collaborations and relationships. Any replacement of this sentiment with a more mechanistic approach is bound to cause upheavals.

5.7.5 *The training requirement*

The contemplation of implementing any change affecting the construction industry concludes in whole or in part with the requirement for training. The subsequent failure of the training programme to materialise usually causes consternation. The case for the training is made out in the following comments:

> I use software and data often – software is only useful when the quality of data is high and therefore the humans inputting the data are trained properly.

The requirement for training and malleable staff interested in the developments are certainly constraints. The investment required would follow from the business case being properly made out and the market leaders setting the pace in this field.

5.7.6 *Computers cannot negotiate*

The first comments discussed seem to resist the move towards automation on the grounds that this is not where the industry has been headed.

> A computer cannot decide this and will only use the facts which in a time of mutual trust and cooperation isn't the way forwards.
>
> Taking the human element out of the contract will not lead to collaboration as the contract will be administered as a binary yes/no process with no option for discussion.
>
> Getting round the table is a key part of collaboration. Can computers argue a point?
>
> Standard construction contracts are "prescriptive" to a point, interpretation and good management makes up a large part of the success of a contract which cannot be written into code.

The alternative view is to bear in mind that computers are a tool and can perform a good deal of the repeatable aspects of construction whilst allowing for human input on the more intractable problems. This is the semi-automated position advocated as likely to be the work around in the short-to-medium term.

5.7.7 *Humans are required for dispute resolution*

Nowhere is the case made out more forcefully in the comments than for human intervention than in the dispute-resolution procedure. Whether or not this ought to be singled out for special mention is debatable.

> Dispute resolution and human intervention seem interlocked to me – how are unforeseen circumstances to be handled?
>
> I believe some straight forward issues can be solved by computers by more complex disagreements with an unclear conclusion will always require human intervention.

If there are any unknown or unquantifiable risk events then a traditional "manual" control that engages all parties to the contract are preferable and advisable.

Taking the human element out of construction is likely to result in a more adversarial approach.

The reduction in risk is always welcome but there is a need for the human brain to be applied to the words.

You can use technology to highlight any problems due to incorrect data, late works etc. This would need face-to-face collaboration to resolve. Can this be done computer to computer?

The last point is again framed as a question to which the answer could conceivably be "yes." The vast majority of the respondents are self-confessed as not being particularly familiar with the intricacies of the subject and presumably are open to the possibility of their being the technologically enhanced answer they question. This is examined in section five.

5.8 The longer view

A few of the respondents made the connection between trust and smart contracts as being capable of residing in the same place.

Trust and collaboration will be automatic if [smart contract] is delivered.

Trust can always be worked upon when something has proven to work well.

Trust will only be gained if the contract is seen as fair and reasonable in the first place.

These insights establish some of the new connections made in the final section of insights.

5.8.1 Testing

It appears to be generally accepted across the respondents that a period of testing and trial and error would be involved before implementation.

As with the change from previous forms of contract to more collaborative workings, any changes, particularly when decisions are taken away from people, systems and processes will need to be tested and bedded in before gaining confidence and trust.

Computers are too glitch and error prone (due to poor information input by human element) computer systems are not reliable enough yet.

There will be many failures before we recognise success.

These points appear incontestable and would, in any event, be part of the process of introduction.

5.8.2 BIM first

The link between BIM and smart contracts has been discussed as being the probable route to implementation. If this is correct then citing comparisons with the BIM journey is valid.

> BIM is still in early days. Long way to go before full automation.
>
> Many of the projects I work on are not fully BIM enabled and I don't think this is unusual. Therefore I think we are a long way off smart contracts.
>
> Much like BIM it is a good idea that will be useful but will take time to integrate into the industry. Projects will become fully BIM enabled eventually.

The general acceptance of where things are going means that the debate centres on the time it will take and the benefits/impacts. BIM appears to be making progress but not, if these respondents are to be believed, at the stellar rates claimed in the National Building Specification (NBS) Surveys.

5.8.3 Next generation

The issue examined here is how respondents judge the passage of time required for something to become established. Erring on the side of a long lead in appears as the default position.

> Always with new technology a critical mass is needed.
>
> I am currently unsure about the future of smart contracting and BIM. I think ultimately it will be positive but it will be many years until it is fully embraced.
>
> It may take a generation to change this as younger staff would embrace this technology easier.
>
> Yes technology is the future for all industries, but smart contracting I don't believe is remotely achievable on complex construction projects any time soon.
>
> The New Engineering Contract (NEC) took 20 years to become the contract of choice and that only due to government intervention so collaboration does not come naturally to the industry.
>
> It will take some time to understand where contractors are able to gain an edge.

This last point can be addressed once the business case is made out. The encouragement appears to be that, as with the NEC experience, real change can eventually be effected.

5.8.4 Priorities for adoption

These perceptive comments appear to establish that: trust needs to come before the business case, the element-based approach is the correct one and that a semi-automated process could be developed using existing contractual procedures.

> Although costs savings are important, I think that smart contracts would need to overcome the trust and collaboration issues first before money were an issue.
>
> Procurement teams rarely understand what it is they are procuring in enough detail and do not want to engage with people to really find the best service or product at a fair price. They just want to set things up through e platform and score against stuff they can understand/measure like whether you have ISO standard. These tell you very little other than some paper work has been filled out correctly. Smart contracts seem like an extension of something that is already not working well.

The last point again resonates with a shortcoming of current procurement procedures. The fallacy that procurers want to know the complete picture before awarding a contract is touched upon here. The de-skilling of the process can be observed as starting much earlier than the construction side of operations. Ensuring the proper application of high-level skill at the key points in the process is as vital as it ever was.

5.8.5 Flipped thinking

The last selection of comments has been chosen for their illumination of counter-intuitive thinking and are analysed independently. The prediction of more disputes is of interest here. This might be along the lines of clash detection in BIM where thousands of clashes are highlighted, only to be resolvable at the move of a mouse click on the design programmes.

> It is the very complexity that means that automation is essential and not a hindrance. Why should a person matter? The facts should.

Computers can manage complexity. The real issue is whether this can be translated into a simple decision on how to the issues to prioritise and arrive at a decision.

> If a smart contract is used as intended then trust and collaboration should be of no real concern as each party's roles and responsibilities would be clearly defined.

This comment reveals the point that trust and collaboration were never the end in themselves but a means to that end which can also be achieved and complimentary to, the smart contract approach. In this comment, the dispassionate nature of the decision-making process is seen as a positive. This contradicts the requirement for the "human touch" espoused earlier. The automation of the judicial process was discussed as a possibility.

Newer QSs are much more "technology native" and do understand the processes which underpin this technology.

This comment supports the Douglas Adams approach:[6]

I've come up with a set of rules that describe our reactions to technologies:
  1 *Anything that is in the world when you're born is normal and ordinary and is just a natural part of the way the world works.*
  2 *Anything that's invented between when you're fifteen and thirty-five is new and exciting and revolutionary and you can probably get a career in it.*
  3 *Anything invented after you're thirty-five is against the natural order of things.*

One of the consolations of being in the third bracket is the ability to take a longer view on the prospects of recently heralded initiatives to embed themselves successfully and to objectively view the obstacles in their path. For younger construction professionals, there is less of a perceived threat and more of an opportunity.

The author's own advice regarding how to navigate a changing work space is to always to back yourself and your professional intuition. Experience in any walk of construction law is hard won and very valuable. This is not to deny technology but it is to recognise it as a tool to complement what we already do rather than as a cause of our redundancy or obsolescence. To our younger colleagues, we can quote the T.S Eliot poem, the Wasteland : "O you who turn the wheel and look to windward, Consider Phlebas, who was once as handsome and tall as you."

5.9 Contributions of solutions to the debate

The priorities for smart contracts and their likely impact in the foreseeable future are summarised below.

In order to overcome problems in the short term it would be useful to:

1 settle upon a definition suitable for the construction industry and its inter-relation with other technologies;
2 form a clearer picture of what it is the industry is currently dissatisfied with;
3 establish which benefits are realisable and by what means;
4 draw the parameters of what is achievable by reference to the limitations which currently curtail performance and how these may fall away in the future;
5 dispel those cynical opinions and fallacies which do not reflect the true picture whilst acknowledging those which are real and present roadblocks;
6 form a strategy for implementation and training, possibly learning from the BIM experience;
7 draw a line under the computers and trust debate by recognising that the two are not mutually exclusive;

8 draw up a realistic timetable and monitoring platform to accompany developments; and

9 take the debate in new directions and recognise the huge potential for positive impact on a dispute-ridden industry.

5.10 Further correlation and corroboration

The McNamara paper takes a more business-headed approach to the potential for smart contracts. The interviews with seven Australian construction industry professionals including barristers and contract software developers. This allowed him to focus on the commercial opportunities presented by smart contracts. A brief review of his findings is presented here.

5.10.1 Weaponised contracts

The contract is used as a weapon when situations present themselves. There are many builders who make more money by not complying with the contract and dealing with the consequences. The industry is tired of this approach with a real need for an alternative contract solution.

We see contracts as an adversarial tool. The only time a contract comes out is if the relationship has broken down and you have to refer to the "rulebook."

5.10.2 Business opportunities

Speed of reaction to any given scenario is key to success for any business. A digitises contract will add an advantage to any "rainy day" situation. Instead of having to trawl through mountains of paperwork which is extremely onerous, having the ability to garner all required information in any given situation will add huge value.

Clients will have more certainty over the delivery of the project through intelligent contracts. Every party has to add contingency to cover risk on a project which carries a $figure. On large projects, this adds up to a huge sum. At the moment the industry is not very investable.

5.11 Conclusions

The views and perceptions presented in this chapter demonstrate the diversity in attitude and opinions that exist throughout the construction sector. There are common themes though and the take-away is that people can certainly see the point and the benefits on offer. The focus is often on the "bumps in the road" in terms of the short-term constraints and obstacles to adoption. In most cases, the respondents display a degree of sangfroid to the future that is coming but recognise they should clamber aboard the wagon, no matter how precarious their hold. After all, the journey might be fun and the wagon is already moving.

Notes

1. https://www.reddit.com/r/Buttcoin/comments/4dcoe1/why_are_smart_contractsdaos_a_bad_idea.
2. Mason, J. (2017) *Intelligent Contracts and the Construction Industry*. Journal of Legal Affairs and Dispute Resolution in Engineering and Construction, 9.
3. Mason, J. & Escott, H. (2018) *Smart Contracts in Construction: Views and Perceptions of Stakeholders*. Proceedings of FIG Conference, Istanbul May 2018.
4. McNamara, A. & Sepasgozar, S. (2020) *Developing a Theoretical Framework for Intelligent Contract* Construction Innovation: Information, Process, Management Special Issue – Industry 4.0 Disrupting the Construction and Civil Engineering Supply Chain.
5. Mosey, D. (2019) *Collaborative Construction Procurement and Improved Value*. John Wiley & Sons.
6. Adams, D. (2012) *The Salmon of Doubt*. Pan Publications.

6 Smart contracts and the legal system

6.1 Introduction

Investors are only usually willing to part with real money with the assurance that there is a legal foundation for their engagement. Thus far, the legal uncertainty that pervades the use of cryptocurrencies and crypto-assets for financial transactions has slowed the rate of their uptake in certain sectors including construction.

A legal foundation to smart contracts is essential if smart contracts are to become mainstream. Investors will need to be able to be able to invoke legal remedies in appropriate circumstances so as to avoid foul play and ensure a dependable market. The smart contract also has the potential for built-in dispute resolution. This is examined in section five. The UK Government took the first steps to establishing a solid legal footing by establishing the LawTech Delivery Panel. The mission of the panel was to demonstrate how English law and the UK jurisdiction can provide a foundation for the development of distributed ledger technology, smart contracts, and artificial intelligence and associated technologies. In relation to smart contracts, the key is to establish whether a smart legal contract is capable of giving rise to binding legal obligations, enforceable in accordance with its terms.

The Delivery Panel's findings were that confidence in smart contracts' legal position can be grown through legal developments and increasing standardisation. The common law provides many advantages including its ability to build on accessible principles of law and apply them to new situations. The key here is the continuity of service – creating a new legal and regulatory regime is discouraged in the Legal Statement[1] that is claimed to be counter-productive to an environment encouraging the new technologies to flourish.

The main advantages of the common law over a more civil law approach is that there is room for relevant concepts to grow organically and for a precedent to be set. The civil laws are left with a more interventionist approach, which can also stimulate confidence and lay the groundwork for innovation. Certain American states have taken this approach; for instance, in Vermont a law has been passed to the effect that[2] *"a fact or record verified through a valid application of blockchain technology is authentic."* If a record on a blockchain is disputed, the party seeking to challenge has the burden of producing evidence to disprove the entry on the blockchain. Interestingly, the identity and ownership of assets recorded on the blockchain are also declared legally valid. However, identity and ownership remain a problem due to uncertainty as to the linkage between parties and pseudonymous transactions without extraneous information.

Another attempt to kick-start legal confidence in smart contracts took place in Arizona where a law was passed to *introduce* basic provisions that declare the authenticity of fundamental aspects of blockchain and smart contracts. The law declared that cryptographic signatures are considered sufficient to act as binding electronic signatures. Secondly, the scope of electronic records was amended to include blockchain. Finally, the use of smart contracts to enforce an agreement between parties was expressly permitted.

> *Smart contracts may come into existence in commerce, a contract relating to a transaction may not be denied legal effect, validity or enforceability because that contract contains a smart contract term.*[3]

However, regulatory uncertainty and shortcomings in the technology regarding smart contracts and blockchain remain.[4] It is one thing for the lawmaker to effectively say, "go ahead and use these mechanisms" and another for the confidence and trust to exist to take the necessary steps. The incidence of trust is considered in section 6.2.

6.2 Informing trust

Smart contracts have been described as the "ultimate automation of trust."[5]

In one analysis, contract law, and therefore the construction law off-shoot, deal in creating sufficient trust for individuals and businesses to trade with each other and thereby allow society and the economy to function. Each party trusts that the other will behave in a certain way. Smart contracts and blockchain are sometimes termed trustless concepts in that the need to trust each other is removed in the traditional sense. There is a complimentary approach, which discusses the need to inform trust.

Trust is informed by a wide range of factors and derived from a variety of different sources. The rule of law, protections of regulators, economic incentives, reputation and brand recognition all contribute to the trust we place in certain interactions. Technology has long played a part in informing trust in interactions as has law through contracts. There appears to be a lack of trust in construction, despite the attempts of the alliancing and partnering lobby to introduce this.

Trust is also built through the deployment of professionals. This is manifested both by the calming reassurance a professional can bring with their expertise and in the accountability which is part and parcel of the appointment. The knowledge that professionals are liable in the event of negligence is a form of protection for the user of their services. The absence of blame in an integrated-project-insurance approach still produces nervousness in many quarters. However, redefining trust may be helpful beyond the usual professional constraints and foresees the nature of trust changing from a fiduciary type duty to something better equated to confidence, which is deemed "quasi-trust." The result sought is that users will find ways of seeking reassurance that what is on offer is reliable. This might come from the provider being a highly reputable brand or a clear endorsement from the government.

This is already seen in consumer transactions where the five-(maximum) star reviewing system for customer feedback provides all the quasi-trust required. The incidence of disputes between vendor and purchaser are quickly settled where the vendor values their feedback as their main bankable commodity as to the quality of their goods and service. The transparency experienced here is linked to the wider movement for openness and collaboration.

The role of quasi-trust is also perceptible in the context of disaggregated tasks that are sourced by providers who are not traditional, mainstream professionals. Quasi-trust can be established in the role of oracles for the very reason that their role is stripped back to the recording or determination of a single issue. An oracle should be able to be sued if it does not perform but our confidence in the simplicity and articulation of the task has to perform gives us trust in it. The Susskind paper refers to this in terms of the demystification of the professionals work. They establish that in this connection, the role of trust becomes of secondary importance and that quasi-trust, backed by a contract or even regulation, will be sufficient.

The most common way of recording data in a distributed ledger in a trusted manner is through oracles – people or businesses tasked with recording specific real-world data into the distributed ledger. The presence of an oracle within a system to some extent re-introduces the need to place trust or quasi-trust in third parties. However, this quasi-trust does not reside in a single entity, as it is possible for multiple oracles, from sensors in tower cranes to cameras on gantries performing multiple functions to exist within a single system. It is this distributed nature of the sensors – as well as the ledger itself – which is the core strength of the technology.

6.3 Smart contract law

The moves towards standardisation in smart contracts emphasis the importance of how they are written, how they are enforced and how to ensure that the automated performance of a smart contract is faithful to the meaning of any relevant contractual documentation.

Different views exist about whether the data revolution really does present the type of challenge that the lawmakers have not seen before. The case of *Miles v Entores*[6] is a good example of the flexible nature of common law stretching to cover a new technology – in that case, the facsimile or fax machine, and how it aligns to notions of offer and acceptance in contract law. The issue was whether the contract was made when the sheet of paper went into the fax machine in Holland or out of it in London. In that case, the communication of the acceptance was the key and the laws of England governed the transaction.

The authors view doubts that the existing law can cope without major statutory intervention. Early on in this book, the theory of the Ptolemaic Epicycles was set out. The contention is that the types of changes predicted are so revolutionary and so much of a departure from the existing that common law will struggle to cope. The disintermediation and peer-to-peer nature of the developments present some of the biggest challenges.

Notwithstanding this, the presumption is that incremental change is the most likely path towards development. It is therefore useful to set out where smart contract are similar to existing provisions. Smart contracts are essentially systems for complying with agreed mutual obligations as to delivery and payment. In order to establish the self-executing transactions of a smart contract, the parties need to express their underlying contractual commitments according to conventional contractual rules.

That is not to say that the lawmakers are blind to the bigger forces at work. Lord Justice Vos recognises this:

> The Technological Revolution is not about replacing books with online legal materials. It is not about using simultaneous transcription of court hearings and digital case management systems. It is not about the increasing use of telephone, video or online hearings. We do all these things already, and will continue their utilisation…It is about the complete transformation of the way in which people will in the future do business and transfer value.

The lawmakers are therefore keen to tackle any fundamental legal impediments identified and align any regulatory regime with what exists. The concern is that a new approach may discourage adoption of technologies further. Another problem in the developing field concerns the fundamental attitude taken. Many of the digital natives reject the conservative approach, which they perceive from such things as judicial pronouncements. And yet, it is these experiences that underpin society. Paying tax is the highest form of civilisation and yet many humans would exploit a loophole to avoid this if they could. The tax evasion gives way to law evasion and that is not acceptable. The pseudonymous nature of online dealing (Nakumoto is him/herself not a real name) has serious implications for the rule of law. Lord Justice Hodge addresses the point thus:

> Our society has the right to demand that technology and technologists are not exempt from or above the law. That is probably obvious to a lawyer, but it is in this generation by no means a given. In order to win the argument, lawyers and legal systems will need to adapt so as to ensure their future relevance to new forms of transaction.[7]

Again, in the interests of balance, it is not necessarily for the judiciary to dictate on peoples' approach and philosophy of business or anything else save as to re-inforce the rule of law as above. The approaches – respect for the rule of law and wanting to decouple from a rigid form of civic engagement and values – are not necessarily in conflict with each other. What the new business may want is the certainty of law without the constraints that might be perceived to come with it. Further, the judges may well have a vested interest in looking for continuity rather than revolution in our legal arrangements. Harking back to one of the main themes of the paper – the temptation for disrupted professions to say "this far and no further" can be illusionary comfort in the fast changing world.

How can the operational side of a smart contract be captured satisfactorily inside existing contract law provision? Nick Szabo used the example of a vending machine to illustrate his concept. A fizzy drinks' machine will deliver the drink if the right coin or other payment is tendered. This displays the traditional logic of contract law – IF A happens THEN B happens. We can see how this might apply to the smart contract in the construction industry – IF the oracle senses that the brick has been properly installed in the wall and the warranty information is provided THEN release the payment agreed. However, what if the wrong fizzy drink comes out or the mechanism is stuck? In the construction example, what if the oracle device malfunctioned or the installer enters insolvency? This is where the interaction with the background law comes into stark relief, as remedies must be available for confidence to exist in the systems.

For some, the shortcomings of the smart contract, in particular its failure to react to situational circumstances, are sufficient to deny that it is actually a contract at all.[8] The argument here concedes that performance may be automated but that a code-based instrument cannot merge with the legal, all-encompassing contract. The legal doctrines which are part and parcel of a contract at law are not present in a smart contract. For example, there is no recourse to the law of mistake, misrepresentation or equitable relief.

The situation arrived at, and the current limit of where we are likely to journey any time soon, Is the stack approach which views smart contracts alongside legal contracts and view both elements as a contractual package, some of which is automated at a performance level (the smart contract) and some of which provide the framework with the background contract law. However, it is intriguing to identify what it is that makes these two parts distinct and whether they can be reconciled into one all-encompassing smart contract. This is the task that is engaging those at the vanguard in their quest to write a computer/smart contract legal language.

6.3.1 Form of agreement

Legal Data Exchange has been around for 40 years. Emails are capable of representing contracts and distributed ledger communication and have the same capabilities. Is there an issue around formation? Not particularly according to Lord Justice Vos – *"new methods of communication do not imply or need to create a new principle or a parallel regime to accommodate online contracting. The revolution in how people communicate need not result in a revision of contract law."*

A contract is a legally binding agreement containing mutuality of obligation (offer and acceptance) definitive terms and a consideration. The term "legally binding" means that the contract is enforceable under law through the legal system. These three characteristics are present in smart contract models. For example, an ERC-721 (Ethereum Reques for Comment) token in Ethereum:

1   An ERC-721 token is offered to the public.
2   Consideration exists where an individual or organisation willingly exchanges a cryptocurrency for the token on offer.
3   Acceptance occurs where the network confirms the transaction and it is propagated across all ledgers.

Provided both parties have entered into the transaction willingly, that is, without undue pressure or duress, the smart contract is enforceable. Users of construction contracts will recognise the above process as akin to a tender situation. The offer of the token is effectively an invitation to tender followed by a tender and the placing of an order through acceptance. There is one departure from the standard rules, which is that the acceptance must flow from the offeree. This feature is effectively delegated to the network to perform the confirmation. The better way to think of this is that it is the performance of the contract that signifies the acceptance by the offeree not its record on the blockchain, which is a later ratification process. This could be viewed in a similar way to the line of cases stretching back to the Carbolic Smoke Ball[9]. This seminal contract law case concerned a flu remedy called the "carbolic smoke ball." The manufacturer advertised that buyers who found it did not work would be awarded £100, a considerable amount of money at the time. The company was found to have been bound by its advertisement, which was construed as an offer which the buyer, by using the smoke ball, accepted, creating a contract. The use of the smoke ball signified acceptance and the requirement to actually communicate this back to the Smoke Ball company was waived by the Judge. The smart contract is executed in accordance with these characteristics.

6.3.2 *Contract interpretation*

General principles of contractual interpretation can be applied to a smart legal contract written wholly, or in part, in a computer code. English law does not normally require any particular form for contract expression. The requirements around formation have to be met and there must be no vitiating factors (for example, duress, misrepresentation or illegality). The rules of consideration must be adhered to, unless the contract is a deed, in that something of benefit must be transferred.

This is not an issue for a smart contract that can usefully be described as penny and the bun transactions. The consideration is either the penny or the bun depending on your standing in this (potentially) smart arrangement. The lawmakers consider it to be evident that a contract composed wholly or partly of a computer code is capable of constituting a legally binding contract and does not consider any new legislation to recognise this. The counter view is that if it would add clarity and confidence to the situation then it would be worthwhile, not least to establish that the United Kingdom is open for business in this regard.

A question which is slightly more involved centres on the behaviour of code in circumstances where it is used to implement agreement but potentially without defining it. The key question here is what the parties intended and whether they intended to be bound by the behaviour of the code. The role of the Judge would be to analyse the words and conducts of the parties in light of the admissible evidence to determine what was agreed.

The Legal Statement on Cryptocurrencies gives the scenario where the parties contract in natural language but arrange for performance to be executed using a code, typically on a distributed ledger. Although this contract involves smart aspects (i.e. code), the contract itself is conventional. Should the code fail to execute as intended, then either the natural language part of the contract will specify the next steps or the Judge would be left with the common law principles to rectify the situation.

The scenario is also envisaged where the contract exists entirely in code. This is where the technology moves furthest from familiar territory. The consideration will still be apparent, as this will be the purpose of the contract. The problem will arise if one of both of the parties allege that the outcome of the contract was not what they had contracted for. The Judge would need to look beyond the four corners of the code to interpret it. This might involve extraneous evidence around intention and mutual understanding. It is in this type of area that precedent and a regulatory framework would be useful. Further complications might involve around whether an agreement has been reached and whether legal relations were intended.

6.3.3 Natural persons

What is the status of the electronic agent? The case of Software solutions V HM Customs & Excise[10] included some observations on contract formation, notwithstanding that the main issue was around tax exemptions. The electronic process of contracting was automated and the finding made that the insurer would be bound by the automatically generated result.

> the fact that acceptance was automatically generated by a computer software cannot in any manner exonerate the defendant from responsibility. It was the defendant's computer system. The defendant programmed the software.

The court drew an analogy with Thornton v Shoe Lane Parking[11] where a ticket-vending machine was held to be an offer capable of acceptance. Even though the contracts were made electronically, rather than mechanically, that did not alter the application of the basic legal principles. The lesson for smart contracting is that the intent of the parties needs to be clearly expressed as to at what point the agreement comes into existence.

Another area causing some concern around the future development of the law around smart contracts involves the use of anonymous and pseudonymous parties. This is not a problem for the law as such – the identity of parties is often hidden in such events such as auctions. The bigger issue is for the other party is around enforceability. The contract may well bebinding but on a party who may be able to avoid liability and on whom the usual credit checks and security for payment cannot be secured.

6.3.4 Jurisdiction

Smart contracts operate via distributed nodes (computers) which may be based all over the world. It will be difficult to determine the applicable governing law and jurisdiction. One way around this in analogous situations involving physical contracts is to include an arbitration clause. This would agree a seat for the arbitration and also set the procedural law which applies. Parties will wish to select a seat which does not render a smart contract illegal or unenforceable, and that the codified arbitration agreement in question will be upheld and enforced by the supervisory courts.

6.3.5 Payment

Smart contracts operating via a blockchain are not tied to traditional currencies such as pounds, dollars or Yen (known as *fiat* money). Instead, they envisage the use of cryptocurrency with all the ensuing risks of volatility. A distributed ledger, on the other hand, can use a fiat currency and separate the smart contract procedures from the public blockchain. Other wider problems may exist in terms of compatibility with statutory regimes for payment. For example, there may be some doubt as to whether notices generated through the use of smart contracts amount to meeting the payment regimes of the Housing Grants Construction and Regeneration Act 1996 (as amended).

The involvement of crypto-assets and crypto-currencies also seem set to revolutionise how we pay for things and where our currency is stored either in a bank or a digital wallet. The movement to make crypto-currencies less of a "wild west" with its accompanying volatility in its market price seems set to continue. Nation states are looking into launching their own crypto-currencies, as are global players in the way of private financiers – JP Morgan have a JPM coin.[12] These result in currencies that are underpinned by a real currency, for example, dollars held in JP Morgan bank. Internationally, countries facing international sanctions embrace the trend for crypto-currencies as this is a means of by-passing another countries' economic sanctions. The availability of cheap electricity and plentiful skilled labour mean that the mining function of a public blockchain is performed for an attractive base cost.

6.3.6 Disputes

Section five looks into the potential for an online dispute-resolution process. Smart contracts will need to address issues such as conditions precedent and the hierarchy of terms within a smart contract, which was partially written in code and partially in natural language. Clarity would also be needed around which rules of interpretation would be used in the case of ambiguity or lack of functionality. This could be the role for a semantic oracle to resolve. The judiciary are probably correct to say that these sort of issues could be resolved on a case-by-case basis and the common law could develop precedent and guidance in an iterative process. As ever, the down side of this is there is uncertainty until such a time as there is certainty through judgments.

The rules of interpretation in situations where there are ambiguities in contract provisions are well developed in English law. Judges usually apply the rules of objectivity and purpose and context to rule accordingly. The literal rule also has judicial following along with the contra proferentem (against the offeror) rule. All of the above rules have their logic in both parties being in a position to understand the contract terms they entered into. This is not necessarily a given for a contract written partially or wholly in computer code. This necessitates a restructuring of the rules of interpretation or the development of a computer language understandable by all.

6.3.7 The position of subcontractors

It is straightforward to envisage how the position of subcontractors would be greatly improved by the use of smart contracts. Implementing the use of smart contracts for managing payment would see each party benefit in terms of predictability and transparency of payment terms. Late payment and being at the mercy of main contractors would be bypassed in the project bank account type arrangements envisaged.

6.3.8 Privacy and criminality

Privacy and criminal uses of the web are excluded from this book. In any discussion about future professional roles, the importance of data security and protection is identified as growth areas. The future role of the digital security guard will be extremely important as well as the law which enables this.

6.3.9 Conclusions

Integration of smart contracts into the norm will probably be long and arduous. There is likely to be a gradual transition through various smart contracting models with a long-term aim of full integration. Laggard behaviour of the construction industry allows the observation of others' experiences and the reactions of early adopters. An appropriate solution can be formulated.

6.4 Smart construction law

Construction law provides a set of rules, which govern the procurement and building of an asset. The ultimate application of smart contracts is to create automatic supply chain transactions where all required decisions have been made and where human judgments and interfaces are required to a much lesser extent. Glimpses of what this future might look like are provided by initiatives such as pre-fabrication and design for manufacture. An interim position will require the "moving parts" of construction contracts to be adjusted rather than replaced. Certification has long been the key to keeping construction projects on time and on budget. The exercise of the professional judgment of the certifier to record the occurrence of an event and to certify the same provides the trust in the system. The contractor is aware that the certifier is ultimately paid for by the client and that this gives an opportunity for a biased approach. However, comfort it taken from the professionalism of the certifier and the ability the contractor has to challenge the decisions through dispute avoidance/resolution. The automation of the certifier role through oracles provides an opportunity for providing a transparent and immutable record of certification. As a result, every member of the network would be able to verify that the certification was issued thereby circumventing the need to trust the word of any bi-lateral contract partner or supplier.

Against this backdrop, it is helpful to focus on the core obligations of the building contracts.

6.4.1 Pay/build

The contract must say when, where and how the Contractor is being paid and what for. In return, the services rendered by the Contractor must be made clear. Advances in cryptocurrency, big data sensors and project bank accounts may lead, at the very least, to the semi-automation of this function. The issue then centres on verification of the completion of the work to the standard required. This function can be made quicker and cheaper by adopting a smart contract approach. Spot-checking and the collation of indicators into a dashboard would represent advancement.

6.4.2 Instruct/obey

The ability to make decisions for and on behalf of the client needs to be included in the contract. The terminology used varies and the office holder may be known as the Architect/Contract Administrator or the Project Manager or Engineer. The extent to which the Contractor is required to obey any direction given needs to be defined. Questions arise around whether the Contractor has a right to challenge any decision made and whether this can be before or after taking the compliant behaviour. The role of the overseer is required, amongst other things, to record when time and money events have happened. Other roles include dealing with discrepancies between contract documents and issuing variations. The overseer has ancillary powers such as excluding parties from site and ordering the cessation of work in the event of force majeure or the discovery of archaeological remains in the area.

The suitability for automated contracts to deal with change and uncertainty is a major issue preventing the realisation of smart contracts. An automated process can deliver the BIM model. Doubt surrounds the ability to code, say, the force majeure clause. Often, there is a list of the type of events giving rise to the possibility of an extension of time and/or extra payment. Discretion is provided for in the wording "or such similar event." This discretion or room for interpretation can be likened to an expansion joint allowing the bridge to move within limits to deal with different pressures and outcomes.

A concern identifed around smart contracts is that they will be unable to cope with certifier discretion type provisions. A computer programme is made up of algorithms which are essentially "if this then this." Can a force majeure clause be reduced to a set of algorithms – can it provide for the unexpected? This, in the long run, is likely to be the major obstacle in the adoption of smart contracts. This is separate from the challenge of having the clients agree to give irrevocable control to the machine. Commentators on smart contracts have talked of the need for a "kill switch" to wrestle back control in the event of an unpalatable outcome.[13] The author's view is that whilst some elements of a construction contract can be automated, such as the PAY BUILD function, there should be an element of human sanction. This is the semi-automated version of the smart contract. Dealing with uncertainty can be interpreted simply as the requirement to provide a dispute-resolution clause. Typically, this refers that in the event of dispute, the parties should refer to either mediation, adjudication, arbitration, litigation or a dispute-review board. However, prevention is better than cure.

The eventual sophistication of the computer in being able to resolve disputes and confront the unexpected is the logical extension of this approach.

6.4.3 Set deadlines/meet deadlines

Timing is a crucial feature in building contracts and is often equated to having the same importance with money. The building contract must therefore provide a mechanism for setting the original timescale but as importantly must also give a mechanism to move the deadline upon the occurrence of certain events. The processing power of existing computers could accomplish the task of predicting and remodelling a programme to take account of eventualities.

6.4.4 Give access/take possession

The Contractor cannot be expected to carry out and complete the works unless he is given access and can take possession of the site. Directions as to how and when this will happen and the responsibilities that pass to the Contractor consequent on this need to be spelt out in the contract. This matching set of key obligations appears to provide no great challenge in terms of automated construction. Provided the necessary safeguards and insurances are in place and the necessary permissions and planning restrictions have been satisfied, automation appears possible here.

6.4.5 *Give design/follow or complete design*

The Contractor must be given some indication of what he is to build. The degree of detail given ranges across the forms of procurements available. Another variable is the extent to which the Contractor is required to complete the design from the stage at which it is handed over. This procedure is used in design and build procurement. Questions can arise around what happens with ambiguities or errors in design information and the interface between different personnel in the design team. Huge advances in BIM have already been seen with the result that automation in this field is far advanced. Some issues remain including inadequacies in the leading BIM protocol, intellectual property issues, collective responsibility and insurance arrangements.

A myriad of other issues exists around building contracts such as insurances, health and safety requirements, sub-contracting, determination provisions and liability for the cost of over-runs. These issues are capable of falling into place once the basic functions are addressed and replaced by automated/smart construction features.

This vision of incremental improvement represents the best chance for smart contract adoption. A system can be put in place between the client and main contractor, and replicated between the subcontractors and suppliers. The payment clause is automatically executed through cryptocurrency once the works have completed and satisfied the contract terms or code. The extent to which human intervention will be required to physically check if an issue is pending resolution will define the degree of automation possible.

Taking an example to demonstrate the potential, a weather provision in a typical construction contract provides some valuable insight. A traditional contract lists exceptionally adverse weather as a relevant event, thereby allowing an extension of time. The contract does not, however, define what is exceptionally adverse; so in theory, this is open to interpretation. However, the New Engineering Contract deals with weather objectively and employs the 1-in-10-year value assessment. This is a term of contract, which can be written into a code and automated. This can be achieved by linking meteorological office recordings against the criteria in the blockchain. This is actively being delivered in a Digital Catapult Research programme known as the Weather Ledger.

As has been noted, creating the algorithms for force majeure and the other instances where matters are referred to the discretion of the certifier is more difficult to imagine. This leads to the position, and the desirability, of maintaining some human involvement in the process. BIM and associated revolutions are disruptive but this gives professionals the choice of evolving new roles for themselves. Operating semi-autonomous construction contracts would appear as such a challenge.

6.5 Conclusion

The idea that there will be smart contracts paying for performance upon the sensors signalling compliance is unlikely to be achievable in a vacuum. There is a link with the range of advances required for the collaborative agenda to be re-imagined for the digital age. The advances in BIM, digital twins, multi-party contracts and project insurance can all be seen as pieces of the jigsaw. The business case for their adoption must remain the focus whilst technology overcomes the temporary barriers of reliability and interoperability. Ultimately, addressing these concerns is a waiting game for the technology to reach the stage in its maturity where it is demonstrably in the public interest and where the industry stakeholders have enough faith in the ability to deliver. The next section considers in more detail these steps in-between and the stages of their development.

Notes

1. Lawtech Delivery Panel (2019) *Statement on Cryptoassets and Smart Contracts*, available at: https://technation.io/about-us/lawtech-panel last accessed 21/05/2020.
2. http://legislature.vermont.gov/bill/status/2016/H.86 8 last accessed 10/07/2020.
3. Arizona Electronic Transactions Act *Arizona HB2417*, available at: https://legiscan.com/AZ/text/HB2417/id/1588180 last accessed 10/07/2020.
4. Gilchrest, A. & Carvahlo, B. (2018) *Smart Contracts: A Boon or Bane for the Legal Profession*, available at: https://www.taylorvinters.com/article/smart-contracts-a-boon-or-bane-for-the-legal-profession last accessed 10/07/2020.
5. Digital Economy Outlook (2015) *Smart Contracts: The Ultimate Automation of Trust?*, available at: *https://fdocuments.net/document/1-smart-contracts-the-ultimate-automation-of-trust.html* last accessed 10/07/2020.
6. *Entores v Miles Far East* [1955] 2 QB 327 Court of Appeal.
7. Lord Hodge at the First Edinburgh FinTech Law Lecture, available at: https://www.supremecourt.uk/docs/speech-190314.pdf last accessed 10/07/2020. The paper starts with the quote, *"Money is fine, Arron, but data is power."*
8. Lim, C., Saw, T. & Sargeant, C. (2016) *Smart Contracts: Bridging the Gap between Expectation and Reality*, Oxford University Business Blog, available at: https://www.law.ox.ac.uk/business-law-blog/blog/2016/07/smart-contracts-bridging-gap-between-expectation-and-reality last accessed 10/07/2020.
9. *Carlill v Carbolic Smoke Ball Company* [1892] EWCA Civ 1.
10. *Software Solutions Partners Ltd, R (on the application of) v HM Customs & Excise* [2007] EWHC 971.
11. *Thornton v Shoe Lane Parking [1971] 2 QB 163.*
12. JP Morgan, available at: https://www.jpmorgan.com/global/news/digital-coin-payments.
13. Kill switch as discussed in articles such as: https://developers.cloudflare.com/distributed-web/ethereum-gateway/kill-switches/ last accessed 10/07/2020.

Section IV

The steps in-between

The previous section forecasted what the digitally-enhanced future might look like and some of the legal challenges this creates. This section reviews how s progress may actually made towards these goals and how might the hurdles encountered be surmounted. To launch a new analogy, the advancements required can be thought of as stepping stones across the river, like in the picture in Figure 7.1. Not every stepping stone in a river crossing is reliable, some are slippery and others give way once weight is applied to them. Nevertheless, with due care and attention, and not a little dexterity and luck, it is possible to arrive safe, sound and dry on the far bank.

The key themes to approaching the steps in-between are transparency, traceability and collaboration. All three attributes are embedded within distributed ledger and smart contract drafting. The extent to which current procurement and legal arrangements promote the three themes is the indicator of the soundness of the step being undertaken. There are limits to the extent to which the attributes may apply and frequently a counter view.

Latham recognised that no construction project is risk free (Figure 7.2). This can be taken a stage further by stating that any project will go well if nothing goes wrong, but the test is whether the project can survive adverse conditions through collaborative risk management.

Transparency has an ethical dimension as well as a practical one. The practical element is the visibility of the process and the trust this builds. An excellent example of this is the project bank account. If the payees are given all of the information on how much and when a payment is made into the bank account then they can form their view on whether their share is proportionate and paid in a timely manner. In the words of the old maxim, knowledge is power. The alternative view is that the commercial approach of businesses and their inter-actions are best left as a matter for them to resolve without the scrutiny of any other parties. This might be termed the "mind your own business" approach.

The ethical dimension of transparency concerns accountability and acting in a proper manner in line with societal norms. One approach is to view every professional decision through the test of whether one would be happy for the reported actions to feature on the front of the following day's newspaper. This high standard of transparency would no doubt cause a degree of nervousness in certain quarters of the construction industry given the observations in section one. The counter view here is that businesses are free to act as they see fit within the limits provided by the law.

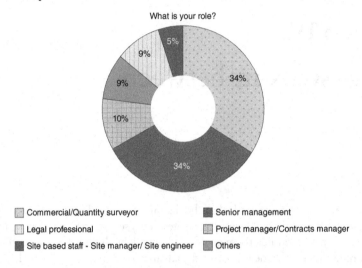

Figure 7.1 The stepping stones towards the TO BE position

Traceability is a less-nuanced attribute whereby the ledger provides the record and it can be scrutinised for its accuracy. The brute force power of data processing enables any entry to be found in virtually no time at all. The limits of the parties' desire to share the information is a factor to resolve in the development of the concepts. The private off-ledger approach appears the most likely solution here where the information is encrypted and shared only within the project eco-system.

Collaboration represents one of the slippiest stone in the river crossing. The extent to which lip-service is paid to the concepts and the limits of self-interest make it so. Nevertheless, working together effectively is undoubtedly required for project success.

Figure 7.2 Principles for progress in construction

7 New collaborative directions

The introduction of smart technologies and distributed ledgers support collaboration. The merging of the two themes of technology and collaboration is writ large in meeting the Government's "Digital Built Britain"[1] vision, which seeks to create a mature digital economy for the built environment, delivering high-performing assets and exceptional client value. Smart, automated contracts can be seen as the logical fulfilment of this desire to go digital. The Government Construction Strategy 2016–2020[2] also encouraged technological advancement in all areas. The Government sought to galvanise change in the area of digital technologies and construction by setting up the Centre for Digital Build Britain 2018. The aim is to deliver a *"smart digital economy for infrastructure and construction for the future."* The Centre moves on from merely promoting collaboration by fully aligning itself with such initiatives as the internet of things, analytics and advances in manufacturing combining to bring about better planned buildings and infrastructure built more cheaply and efficiently.

Repeat business has long been the cornerstone of collaboration. Assembling a like-minded set of professionals and contractors who can work together effectively is the goal. There appears little sense therefore in breaking up that team and running the risk of dysfunction on subsequent projects. The logical extension to keeping a team or an approach the same is to learn from what went well and what did not via data analytics on the project. This has been labelled "enterprise contracting."[3] The focus is on incrementally increasing the quality and quantity of information shared between the participants and to deal with contingencies before they arise.

A quarter of a century on from Sir John Egan's call for collaboration to be the norm, it continues to be treated with suspicion in some quarters of the industry. Those that embrace this approach rarely wish to contemplate a return to the alternative ways of working including competitively tendered work with one-off relationships. However, the doubt exists as to whether commitment to partnering at all levels is possible or even desirable. A misgiving around the partnering agenda is whether it works as well in a recession as it does in a positive market. Certainly, the anecdotal evidence from the last recession was that many employers were very quick to turn their back on the progressive approach and exploit their dominant market position in seeking ever-lower competitive prices from the providers.

7.1 The challenge for the law makers

The role of law in relation to these new developments is to facilitate, accommodate and protect. The developments and trends studied in the last two decades have centred on partnering arrangements and issues such as the concretising of the duty of good faith. The stumbling block preventing these initiatives from taking a firm hold has been the intransigence of the key players in the construction industry and an all too common reference to human nature being against anything resembling a cultural shift away from the distrustful approach which is regrettably the default setting for many.

Part of the role of law in seeking to facilitate and protect the delivery of new initiatives is of central importance. Law and lawyers are often portrayed as being conservative and reactionary in character. This is not always the case however and legal acceptance and verification of new initiatives is a vitally important step in the establishment and bedding down of new practices into the realms of sound practice. The role of the law is to validate the means by which new ideas are delivered and to ensure that they are fit for purpose and that the stakeholders concerned have had the opportunity to check their own position with regards the new departure. Law can provide the 360-degree review of the new idea and ultimately translate the same into concrete and reliable laws. Law is a necessary portal through which a new development must pass in order to attract confidence. This is true at least until a new portal is created.

Law-making can also be a reaction to a situation experienced. The Grenfell fire tragedy is an example and the extremely useful and insightful input from the Hackitt Review of 2018 has already been discussed. Prosecutions are potentially pending here and the final deliberations on individual culpability are still being investigated. Certainly, procurement decisions based on price alone will attract considerable negative attention in the future.

The passage of new initiatives into law is rarely a smooth process. This is accentuated when the changes required are as fundamental as those envisaged in the current fourth industrial revolution are. Many are the Bills of Parliament that have not made it into law because of a failure to find sufficient parliamentary time to conduct the labyrinth of committee stages and readings required in both lower and upper chambers. A streamlining of these processes is probably long overdue and therefore the law makers should be encouraged to look at their own procedures and benefit from digital collaborative working.

There is frequently also a choice as to whether or not to embark upon a new legal direction. This "no thanks" approach was seen in response to the Housing Grants Construction and Regeneration Act 1996 where the oil and gas sectors lobbied for an exemption to the law and were duly allowed to opt out. Other opt outs include the Contracts Rights of Third Parties Act 1999 which many of the standard form contract writers studiously avoid by contracting out. The protection of vested interests can manifest itself in challenges to the law and can result in a seemingly workable initiative might not culminate in an accepted development. Law is extremely practical in this analysis – if a new idea does not translate into real-world interests and behaviour then it is unlikely to become established. Certainty and clarity are the values that law makers and contract writers prioritise above others.

In chapter 2.5–2.6, the patchy track record of government intervention and not being good with change were addressed. This section seeks to chart the positive movements establishing themselves in the development of the agenda for change. The limits of the collaborative agenda are discussed whilst recognising the important role they have to play in providing the smart contract infrastructure discussed in section three. There is also coverage of some of the central notions involved and the enactments in law required to accommodate them. The concept of good faith is a case in point where we can see an example of where a widely accepted positive development in the law comes up against the requirement for clarity and certainty. This section also investigates how the law can keep abreast of some of the technological developments in law both current and hinted at for future consideration.

It is difficult to put a positive spin on the construction industry's position as being amongst the latest of adopters of new technology. Other industries have successfully overcome the inertia generated by the traditionalists so why not construction? It is difficult to see what prevents similar progress being made. The only plausible explanation for this poor position, and one which translates into a truly different set of circumstances to any other industry, is the multiplicity of stakeholders and the one-off nature of their involvement. This factor can prevent collaboration unless a strategic approach is taken to overcome the issues presented. Manufacturing and process-driven industries need only satisfy their client and government stipulations. In the built environment, many different-interested parties are involved. Taking a residential block of housing for example, the stakeholders involved are not just the financing client but also the initial users of the building and the subsequent owners. People feel a strong link to their built environment and the effect it can have on their daily existence. Construction and maintenance of building tend therefore to generate strong feelings that can be difficult to reconcile with the capital appreciation targets of the builder/financier.

The reconciliation of these interests can be facilitated by technology. The required combination of technology and law can ultimately lead to collaboration. In this analysis, it is legitimate to claim that collaboration is "baked into" the technological and legal arrangement that now lie a step or two distant.

The challenge for the law makers is thus to select the right stepping stones for the construction industry's benefit as a whole. In recommending forward motion, the following justification of the selection of the steps in-between is based on novelty, intuition, scalability, alignment, iterative processes, accessibility and status as being worthwhile. Taking these in turn:

- Novelty – there should be some novel aspect to the undertaking beyond the fulfilment of a common sense development. Partnering has partially suffered from this lack of novelty. The argument is that we are collaborating in our own ways every day and this is nothing new for the willing.

- Intuition – based on experience of what works and what does not, does this initiative convey the sense that it is practical and can the result of a working process be envisaged without fundamental problems? For example, a short explanation of the nature and point of smart contracts usually elicits a positive response as the interrogator "gets it."
- Scalability – will this initiative translate to a typical construction project or is it likely to be preserve of mega projects and the public sector only? For example, the Framework Alliance Contract (FAC-1) requires a considerable throughput of work to make it viable.
- Alignment – does this initiative align with other strategies and can it be seen as being complementary or providing an enhancement? For example, a project bank account approach resonates with the other technological advancements being advocated.
- Incremental – are stages in the development of this initiative apparent and is there the sense of building on achievements on that journey? For example, building information modelling (BIM) can be viewed as an enabler for the later technologies to use.
- Accessibility – is the initiative likely to be hidden behind a pay wall or not be generally available? For example, integrated project insurance is not a product offered by insurance companies on the current market has a case to answer here.
- Worthwhile – ultimately, is this initiative fundamentally workable and in sufficient parties' interest, including societal interests, to be worth the effort? The initiative needs to fulfil the role of a larger agenda such as sustainability, resilience or the promotion of a circular economy.

These justifications provide a useful set of tests in the discussion of the merits of the respective steps in-between and reference is made to them as appropriate throughout this section.

7.2 An evolved concept of collaboration

Defining partnering is difficult. It has proven to be an elusive concept to grasp. Partnering is usually indicative of a process involving collaboration and sometimes the use of an open-book arrangement where there is some degree of transparency in the Contractor's dealings with the Employer.

The philosophy of partnering is dispute prevention, conflict resolution and equitable risk allocation rather than the legalistic and confrontational approach taken by many in the industry. Partnering also involved the alignment of values and working practices by all supply chain members to meet client objectives. This is achieved by a shift of emphasis from hard issues such as price to softer issues – attitudes, culture, commitment and capability. The process usually starts with kick-off workshops to establish mutual objectives.

One view on the topic of collaboration promotes the crucial test "is there a *willingness of a party to be vulnerable to the actions of another party based on positive expectations regarding the other party's motivation and/or behaviour.*"[4] This observation seems, at first, to fall foul of the intuition justification for new steps. Who, but an exceptionally altruistic person, would be willing to be vulnerable to the shortcomings of others? This leads on to a consideration of an evolved concept of partnering based on co-opetition. Co-opetition was a term first coined in 1913 meaning co-operative competition based on the principles of fair competition, equality, co-existence and development.[5]

The core principle at the heart of new collaborative directions ought to be therefore that the client can access the benefits of supply chain integration without necessarily becoming too involved in the early exchanges and therefore letting the suppliers get "too cosy." This can be seen in the co-opetition approach and in the distancing of the main contractor from its pet (or domestic) subcontractors. The same intuitive idea applies here as elsewhere in the technological improvements – strip out the hidden costs and non-transparent activity. The online market place of Amazon gives a good example of co-opetition. Amazon lists competitors' products beside its own knowing that it will secure enough business from those who select its products. It will also earn a commission on the orders placed with its competitors as well.

The move towards uncoupling main contractors from their sub-contractors is another instance of potential disintermediation and reintermediation requiring a greater role as a facilitator of the specialists' roles. There should be work enough in this scenario as the main contractor is themselves into the role of the auditor or oracle.

Virtual reality can also accommodate the auditing role from the client side by allowing the client to visit their new building and decide on the desired layout and visualise this. However, it is the automation provided by augmented reality where the real breakthrough arguably lies. Augmented reality overlays actual with planned through the medium of immersive technology and eye wear/glasses. On a building site, this can easily be embedded in the personal protective equipment; for example, Microsoft HoloLens provides the hard hat and eye wear needed for this purpose. When planned is overlaid on actual both before, and more importantly, once built, then checking against the plans (the 3d model) becomes much more straightforward. Take, for example, pipework in an office development. Viewing the installed pipework and connectors whilst overlaying your vision with an image of how it should look makes auditing the work very straightforward. In this instance, the role performed by the technology is not so much an aid to the quality checker as a game changer in the work they carry out. Quality checks become by exception-only, the question is asked, "is there any part of this installation that does not conform exactly to the plans?" Ultimately, of course, the auditing itself can be performed by the computer system and the human involvement goes down another level. Data facilitates this development.

Augmented reality can therefore become the embodiment of the collaborative approach. The downside of BIM is that it is repetitive rather than additive. Further, there is currently the unhelpful habit of giving all the design information to everyone and asking them to pick through what they need for their input whilst making them liable for any omissions or incompatibilities in other parts of the design. Quite often, all the contributor wants is a single piece of information rather than the whole package given to them. Taking contracts back to a simple transaction basis – put that there and I will pay you – will surely help. Using augmented reality technology to see exactly what dimensions and constraints exist should help deliver that item of the build.

A co-opetition approach should help address some of the counter-intuitive downsides of the old collaboration model. The limits of the partnering trust equation are not hard to find when the parties speak frankly and candidly. Clients lament the lack of competition in favour of inflated costs and contractors bemoan the limited budgets and inadequate costplans.The middle ground between the supporters and detractors of collaboration is therefore co-opetition, where the project has aligned properly with best practice and can exist in a well-functioning procurement route that is ideally suited to ongoing technological enhancement and breakthroughs as a platform for the data-rich environment to follow. The consortia based procurement approach also contains this philosophyby encouraging collaboration amongst the participants whilst permitting the client to remain contractually distant from it until ready to enter the agreement.

7.3 Recognising the importance of maintenance

Operation, repair and maintenance are fundamental to the use, safety and financial viability of all completed projects. All too often, these aspects are overlooked in a discussion around construction with the focus solely on the building process itself. However, it is the capability of the construction industry to integrate the whole-life approach that represents an opportunity for real tangible rewards from the use of technology.

A term contract often governs the repair, maintenance and operation of a completed project. A term maintenance contract creates a system for ordering scheduled type of works, services or supplies. As such, it is ideally suited for the smart contract approach. The throughput of work is present to generate sufficient interest from the contractors in terms of regular and repetitive work. It can also govern the call-off of goods, materials, equipment and any other capital project components manufactured off site. The required activities vary widely in complexity and scope and operational providers are aware that clients are particularly attentive to measures that seek to minimise their scope and costs.

Effective maintenance of a building requires auditable, transparent and collaborative processes. Moves towards the automation of these functions reflects well on the wish list of justifications including scalability, intuitive and alignment. The key here is the provision of a continuous service and the need for a golden thread of good quality information. As previously remarked upon, it is all about the data. The maintenance of the building therefore starts much earlier than the completion of the physical works and needs embedding in the preconstruction process in the build-up of a BIM platform. Sharing data at this stage can lead to intuitive development of the maintenance scheme for the building and creative thinking, helpful actions and fair decisions are enabled.

Facilities management (FM) provides a test bed for smart contracts given the lack of criticality in time considerations when compared to a complete project build. FM work is also repetitive and scalable which suits rudimentary smart contract applications.

7.4 Multi-party arrangements

In the United Kingdom, the standard form contract providers stand at a crossroads of accepting multi-party contracts as the way forward. Some standard form writers have embraced the concept of multiple parties much more readily than others have. The first multi-party contract of the last twenty years was the Project Partnering Contract (PPC 2000), which has proved popular amongst its following and provided the approah for the more recent FAC-1. Other writers have taken a more cautious approach such as providing freestanding non-binding charters for single projects. The Joint Contract Tribunal (JCT) Partnering Charter is an example of this. Another approach is to use "bolt on" clauses for standard agreements. This was the approach of the New Engineering Contract (NEC) until 2018 when the next step was taken by the NEC to provide a "true" alliance arrangement with the client and all key members of the supply chain called "partners" engaged under a single contract.

Multi-party contracts are designed for use on major projects or programmes of work where longer-term collaborative ways or working are to be created. Further collaboration can be achieved through the use of secondary option ×12 to augment the partnering credentials by setting common objectives for the project. The JCT also provides popular frame work "umbrella" type agreements which are compatible with their standard forms.

Many of the limitations referred to in section one – disputomania and the insolvency risk amongst them – are arguably a result of the linear contracting arrangement widely used in the construction industry. One of the key characteristics of this approach is that the client is removed from the subcontractors and suppliers providing the services and has no visibility on their financial health or involvement with the design.

The common purpose sought by those taking a more enlightened approach is far easier to generate when a hub-and-spoke model is employed and that is what BIM and collaborative approaches have moved towards. Richard Saxon's quote was that what partnering needed to succeed was BIM and this is self-evidently true. The hub-and-spoke model reflects the multi-party approach discussed above. However, in time, the "hub and spoke" model may give way to the data driven stigmergic interconnected approach. A distributed ledger can facilitate the record of the multiple transactions undertaken and smart contract formation and execution imagined in its conception. One might think of this a project being made up of thousands of mini-contracts all being entered into, executed and paid for with

Linear   Hub and Spoke   Stigmeric

Figure 7.3 Contract types

a greater reliance of automation both in replacing labour and the professional roles involved (Figure 7.3).

A useful starting point is to depict the hub and spoke model as a stage in-between linear and stigmergic. This allows reflection on its benefits over linear, and also its shortcomings when compared to stigmergic (Table 7.1). As can be observed, the stigmergic approach is itself multi-party, on the one hand, whilst also being bi-lateral, on the other hand. Ultimately, this may represent the most positive arrangement if the legal infrastructure were able to confer the confidence required to use it.

The extent to which the stigmergic approach embeds the collaborative approach is debatable. In the author's view, digital technology can substitute collaboration and will stand-alone once the benefits of collaboration are baked-in to its formula.

As identified, smart contracts do not necessarily need to be multi-party in the same way as partnering contracts. There are obvious benefits to having a single programme or contract with multiple users each of which can input and extract goods in return for remuneration. However, strictly, the stigmergic approach shown in Figure 7.3 can also be a whole series of bi-lateral contracts and not therefore multi-party at all.

The counter-argument is that smart contracts do not need to align themselves so closely with the BIM agenda and can return instead to a simple transactional basis. This would result in not one multi-party contract but thousands of straight-forward contracts executed by performance and self-executing. This approach could be facilitated by the adoption of cryptocurrencies in the construction context. The interim payment for component parts of a build can use the blockchain technology described. Each component would be individually chipped and once big data sensors attest to its successful installation and function then the payment will be generated. Human intervention here is not strictly required. The simpler the construction or engineering component being undertaken, and the shorter duration, the better in the first instance. Laying rails or achieving electrification of a line could be relatively simply ascertained.

The scenarios described above have all focused on what happens when things go according to plan. One of the most frequently asked questions by construction lawyers is to ask whether due consideration has been given to what happens when things go wrong. This question is addressed in the next section.

Table 7.1 The relative merits of contractual approaches

Linear limitations	Improvements in multi-party	Enhancement in stigmergic
The bi-lateral nature of contracts mitigates against clear roles and may contain liability gaps/duplications.	Each party sees the roles and responsibilities of the other alliance members and can check so as to check there are no gaps and overlaps.	Exact definition of scope of works and each actors means of performance.
Frequently the main/ sub-contract terms are not shared with other suppliers. Frequent imposition of harsh terms on those with lesser bargaining power.	There is transparency regarding the contract conditions as everyone is engaged on the same terms.	Smart contract code is transparent and available.
The integration of a subcontractor programme on the main programme of works can create issues.	There are single integrated programmes for delivery and deadlines.	The inchstone nature of smart contracts makes timeframes and deadlines that much more granular and deliverable.
Separate professional sign-offs for the component parts of the design can lead to issues.	There is a single governance structure which can be represented by an organogram or similar.	The fulfilment of a smart contract task is audited by the oracle arrangement and recorded in the blockchain.
Design rights and collateral warranty assignment can cause issues.	There are reciprocal rights granted such as direct mutual Intellectual Property licences.	These reciprocal rights would be contained in the smart contract code.
Risk dumping on the unsuspecting can place design risk with those least able to handle it.	Although depicted as a hub and spoke, there are likely to be groupings within the structure reflecting peer support and mutual reliance without routing all contractual concerns to all other partners or via the client as an intermediary.	The distributed nature of the information and single source of truth in the live time smart contract can ensure all parties are on the same digital page.
The certifier may have a conflicting view to the contractor that is not helped by the absence of a common contract.	The certifier is no longer separate from the main contract. The certifier becomes integrated in a way they are not in a linear contract arrangement thus making much more accessible and relatable within the contractual framework.	The oracle role sees the certifier role become less pressured in that there are data records available of the as built asset on which to resolve any dispute arising.

7.5 Multi-party liability

The PPC 2000 and the FAC-1 are both multi-party contracts, which the main participants sign thereby undertaking contractual duties vis-à-vis each other. This has resulted in nervousness amongst some legal commentators as to the position of the Contractor who fears they may be at risk of receiving claims in contract from more than one party in respect of the same circumstances. This situation is known as "double jeopardy." A situation may arise where the Contractor causes delay on a project to find liability not only to the Employer but any other party who can establish that they too have suffered loss as a result of the Contractor's default. For some, there appears to be insufficient clarity about the standing of these clauses and concerned about lawyer's inability to advise their clients on the exact meaning of the terms in the contract. These lawyers will point at a cloudy picture in which it is not possible to identify the party responsible for each contribution and the direct or indirect losses it has occasioned. Some lawyers subsequently advise their clients to avoid these contracts.

The problem is seemingly exacerbated once a BIM platform is used and there is an even greater mixing up of inputs and amendments making it harder to trace the originator or contributor to a problem.

There are two main solutions to the multi-party liability issue, as discussed below.

7.5.1 No blame approach

To enter into a truly collaborative "alliancing" contract, parties arguably have to be brave enough to take the leap of faith that they cannot generally sue each other if matters go wrong. No blame clauses provide a contractual method of doing this by limiting the parties' rights of action to being able to claim for the cost of the works. The philosophy is that it is possible to cultivate more efficient and innovative practices if collaborative working cannot be undermined by the fear of claims. Doubts remain as to whether parties can really leave behind the typical mind-set of adversarialism.

Scratch a little deeper, and a number of exceptions exist where liability remain notwithstanding the "no blame" tag. Fraud, statutory requirements and non-payment are standard examples. These carve outs can create a strange hybrid of liability and no liability. The risk is that seeking to redefine where liability does and does not arise can create uncertainty. Take for example the "wilful default" exception on the NEC 4 Alliance contract. This concept appears to ignore the law of negligence that was itself a solution, honed over the decades to plug the liability gap where professionals (and anyone else) fall below the standard of which others are entitled to expect. Other carve-out exceptions include indemnities for infringement of intellectual property rights and failure to take out insurance. The effect of this carve-out is to dilute the collaboration and "no blame" approach to the extent that the contract resembles a more typical, adversarial-style contract along with its accompanying fraught and ill-tempered negotiations.

In the author's view, the "no blame" approach is an over-simplification of a situation which is not fully understood by those seeking to remove the risk of litigation. Law's purpose, as has been seen, is to facilitate and provide a framework in which parties can operate with the security and safety of their mutual obligations. The no blame clause therefore falls foul of the justifications in terms of intuition. The pro-rata approach is therefore preferred for the reasons stated below.

7.5.2 Pro-rata approach

The fear voiced above that mistakes made on a multi-party arrangement are untraceable is probably unfounded in most cases. The preferred solution to multi-party liability is therefore to use a pro-rata clause. The example below is taken from the FAC-1:

> any alliance member who contributes or prepares a document shall be responsible for the consequences of any error or omission in, or any discrepancy except to the extent of its reliance on any contribution provided, by any one of more other team member.[6]

This is a neat contractual device save to the extent that it might create a circular argument around the original cause of the problem – who relied on whose mistake and to what degree was this compounded by later actions or omissions?

In a further development, the PPC 2000 has an option to change the pro-rata blame clause into fixed shares.

The inspiration for the pro-rata or proportionate approach was doubtless the collateral warranty arrangement for a net contribution clause. This mechanism provided the warranty provider with a defence against being found liable for issues where other professionals or providers were also to blame, to whatever extent, for the loss occasioned. The collateral warranty version of the clause created an assumption that the other parties responsible had paid their share of the claim and that the warrantor could not be pursued for the same.

The pro rata clause is preferred here not because of its unambiguous but instead due to its compatibility with another of the developments featured in this book, namely the oracle role. Oracles, and their limitations, have been introduced as fact checkers and auditors as well as in their potential role as dispute resolvers and arbiters. If the issue of accountability was difficult to untangle in a shared BIM platform with multiple inputs and embellishments on an original design then the oracle could take a view on the issue. The parties would have pre-agreed to respect the decision subject to their right to appeal.

Examining the basis of liability for multi-party contracts provides an interesting example of where lawyers will continue to have nervousness and respond with efforts made to provide re-assurance. The point is not to stifle innovation which represents welcome new thinking. However blanket measures such as no-blame clauses may not represent the surest path to advancement in line with compatibility with existing arrangements and workable solutions.

Chapter 8 takes this point further in seeking to assess which elements of the current infrastructure around construction contracts provide support for the more important developments identified in chapter 9.

Notes

1. Digital Built Britain, available at: http://www.digital-built-britain.com last accessed 10/07/2020.
2. Government Construction Strategy 2016–2020, available at: https://www.gov.uk/government/../Government_Construction_Strategy_last accessed 10/07/2020.
3. Mosey, D. (2019) *Collaborative Construction Procurement and Improved Value.* John Wiley & Sons, USA.
4. Malhorta, D. & Lumineau, F. (2011) *Trust and Collaboration in the Aftermath of Conflict.* Academy of Management Journal 54 (5): 981–998.
5. Williamson, O. (1979) *Transaction Cost Economics: The Governance of Contractual Relations.* Journal of Law & Economics 22 (2): 223.

8 Background provisions

The discussion in this chapter is around the enabling provisions which form the fabric supporting the more dynamic agents for change to be identified. These developments have been useful in signposting the highlighted path towards adoption of later technologies to be found. In many cases, they are part and parcel of the characteristics that a mature industry in the early 21st century ought to display. The initiatives chronicled here have fulfilled a purpose in changing attitudes and bringing about progression towards collaboration. However, their non-inclusion as dynamic agents recognises that they do not satisfy all the seven facets of the test of usefulness – novelty, intuition, scalability, alignment, incremental, accessible and worthwhile – and yet, their contribution is positive and worthy of recognition.

In a typical construction law text, one could now expect a discussion on the standard forms of contract available. This has, to a large extent, already been covered in the accompanying book, *Construction Law, From Beginner to Practitioner*, and reference should be made to that work. A discussion on the newer standard forms appears in chapter 9. The central role of the standard forms in promoting dynamic change is considerable, provided that the philosophy of the approach translates into actions by the stakeholders involved.

8.1 Latham's wish list revisited

It is helpful to return to the Latham report of 1994 to examine the wish list of what modern procurement and contract arrangement needs to achieve. In most cases, this wish list has been fulfilled and more progress has been made than originally envisaged. The statement is largely true; therefore, modern contracts address the needs articulated. Twelve of the thirteen Latham's requirements are discussed here. This discussion is based on an academic paper.[1]

8.1.1 Duty of fair dealing with all parties

A specific duty for all parties to deal fairly with each other, and with their subcontractors, specialists and suppliers, in an atmosphere of mutual co-operation.

The good faith clause in a contract tends to divide opinion between the supporters and detractors. One the one hand, it can be viewed as beneficial to creating a collaborative culture and at least does not do any harm to include it. On the other hand, it can be viewed as a dangerously uncertain clause, which is open to massively different subjective interpretations and, as such, is more trouble than it is worth.

A 2020 paper on the subject[2] states that despite increasing court recognition of express good faith clauses, there remains uncertainty as to whether good faith can be implied into a contract and, with respect to express of implied good faith, how to interpret and apply such a duty. According to Mosey and Jackson's paper:

> Some judges are comfortable relying on such clauses and perceive that they give rise to tangible obligations, all the more so when these clauses support other more specific obligations. Others have made it clear that [they] are unlikely to cut across or qualify other express obligations.

The following cases give an example of two different approaches:

The first case is *Wilmot Dixon Housing Limited v Newlon Housing Trust*.[3] This was a case involving the Project Partnering Contract (PPC 2000), which contains, at clause 1.3, the obligation to work "in the spirit of trust, fairness and mutual co-operation." The issue was whether this obligation covered the dispute-resolution provisions of the contract and wider statutory regime. The Claimant commenced two adjudications against the Defendant in which they were awarded a decision they then sought to enforce. The Defendant resisted enforcement on the basis that the original Referral(s) had been improperly served. The Judge found that the failure of the Defendant to raise the issue of proper service with the Claimant was in breach of its stated intention under clause 1.3. The Judge therefore allowed enforcement. Viewed objectively, this does seem to be a correct approach given that prolonging the dispute by obliging the parties to start over was not likely to bring them any closer to a lasting resolution.

The second case, *TSG Building Services Plc v South Anglia Housing Limited*,[4] addressed the question under the Term Partnering Contract as to whether the termination clause had to be effected in good faith or at least reasonably. His Honour Judge, Akenhead, held that clause 1.1 requiring the partnering team "to work together and individually in the spirit of trust, fairness and mutual co-operation" did not require the Defendant to act reasonably in terminating the contract. The clause entitled termination for any reason or even no reason. The point that emerges from the decision is that a clear and unqualified right in a contract was not intended to be displaced by a general good faith obligation.

Reconciling the two cases is difficult and results in a continued nervousness around good faith clauses amongst construction lawyers. The position remains therefore that the England and Welsh jurisdictions should rely on the other remedies at their disposal to deal with unfairness as per Lord Justice Bingham's direction in the leading case of *Interfoto Picture Library v Stiletto*.[5]

A particular risk of good faith clauses centres on using a clause in a separate overarching agreement, which can be at odds with the underlying terms. A twin-track contractual structure can make it *"very difficult to rationalise the different terms among different contracts and inconsistency among the contracts is almost guaranteed."*[6]

It is predicted that the technologically enhanced future will not need good faith clauses. This is probably wise as they would be challenging to code into smart contract language given their subjective nature. In adopting the granular transactional approach of smart contracts, the state of a person's faith, good or otherwise, will not matter as it should again be "baked in."

That said, a wider problem for smart contracts is how to incorporate the other remedies, which are usually available to contracting parties into the agreement. Estoppel and equitable remedies are part and parcel of contract law, which would not be replicated in code, certainly any time soon. This re-inforces the submission that smart contracts will remain as having some automated performance and payment provisions only for the foreseeable future.

Another issue is raised on the international nature of smart contracts. Other jurisdictions have embraced good faith and have recourse to it regularly. For these smart contract writers, it may be inconceivable to leave out such provisions. They may wish to refer any good faith issue to an oracle for a decision on whether the highlighted action is compatible or otherwise. The wider issue is how smart contracts navigate the choppy waters of jurisdiction and applicable law.

8.1.2 Teamwork and win-win solutions

Firm duties of teamwork, with shared financial motivation to pursue those objectives. These should involve a general presumption to achieve "win-win" solutions to problems which may arise during the course of the project.

This requirement chimes with the common-purpose directive and the benefits anticipated from tactical, if not strategic, collaborations. The optimum situation is to have the two circles of self-interest overlap as much as possible. They can never completely overlap given their existence as business responsible to shareholders and with a purpose to maximise profits. They can certainly overlap more than they currently do, and progress has been made towards this goal through framework agreements. A joint venture or a private finance initiative project also achieves a common purpose.

There is concern that written processes and contracts may stifle collaborative relationships and actually inhibit innovation. Creating and maintaining the right balance between clear rules and spontaneous collaboration is a delicate balance to achieve. However, the timely exchange of information and data and any devices that facilitate this can only improve their chances of success.

The commitment to a common purpose is also seen in other industries such as aviation where Rolls-Royce had a philosophy of selling "power by the hour" rather than selling engines and maintenance packages to their customers. All the customer wants is his airplanes to be in the air with the right performing engines and enabling them to get on with their business. This outcome-based approach is very different to the construction industry where the participants are much more interested in the inputs – the cost of the materials and the labours with their accompanying profit margins. The profit margins are often ridiculously small which brings with it its own tensions. Taking the outcome-based approach is another enabler of a technologically enhanced future. If the procurement and construction of a building is judged in terms of its outcome then many of the issues in the list above fall away. The alignment of the common interest becomes that much more achievable.

The common purpose will therefore be maximised by the adoption of the outcome-based approach to construction as facilitated by greater automation and technology. This is much more readily achievable if the technology allows the parties to know exactly where they are with regards their interactions.

8.1.3 Integrated package of documents

A wholly interrelated package of documents which clearly defines the roles and duties of all involved, and which is suitable for all types of project and for any procurement route…standard tender documents and bonds would also be desirable.

All of the standard from contracts take a "suite" approach in order to alleviate the risk of a twin-track contractual structure. Such arrangements still exist where some framework contracts are used as an overlay on project-specific contracts and can crate serious risk. The flaw in using separate contracts is that in practice it is very difficult to rationalise the different terms among different contracts and inconsistency among the contracts may well ensue.

The recent extensions of the contracts contained in the New Engineering Contract (NEC) suite to include sub consultants and suppliers show the desire for further integration at all levels. Similarly, the FAC-1 form of contract seeks to be agnostic on the selection of the project-specific contract below it. The time when there was the risk of a poor interface between contracts therefore appears to have passed. Again, what mature industry in the 21st century would not have found a solution here to at least make its contracts compatible? This ought to have been a fait accompli in contractual development and has become so.

8.1.4 Simple language and guidance notes

Easily comprehensible language and Guidance Notes.

All contracts now provide guidance notes, and have put a considerable amount of work in making their language clear. The simplicity of the language becomes much more important when one considers the current undertaking of making the legal obligations readable as code and vice-versa. The writers of the new legal/computer language will have to follow the KISS principle – Keep it Simple Stupid – if they are to achieve a middle ground where both have a complete understanding of the provisions involved.

8.1.5 Role separation

Separation of the roles of contract administrator, project or lead manager and adjudicator.

The development of contracts towards multi-party arrangements presents the opportunity to remove the limitations of the Architect/Contract Administrator role. In the absence of a direct multi-party contractual relationship, the liability is limited. Hitherto, a contractor had no recourse against the certifier given that the latter's contractual relationship was with the client alone who would not suffer any loss from the under-certification of the Architect. The closer contractual nexus in a multi-party contract means, in theory, that the contractor would now have such recourse. Admittedly, to be discussing contractor/certifier litigation is to miss the point of collaborative contracting. However, the closer proximity means that there are unlikely to be situations where the client make decisions based on conflicting views and conflicting information provided under different contracts. Latham may have foreseen this occurrence by advocating the separation of the project manager role from the designer's role.

The creation of new roles in part of the re-intermediation process discussed earlier. New opportunities such as the BIM Manager and the rooftop surveyor bear testament to the resourcefulness of the professions and their desire to stake out new territories. Elsewhere, other professionals voluntarily cede their territory as is the case with architects and the contract administrator role (see chapter 1.4).

8.1.6 Risk allocation

A choice of allocation of risks, to be decided as appropriate to each project but then allocated to the party best able to manage, estimate and carry the risk.

This restatement of Abrahamson's Principles re-enforces their status as the fundamental theory of building contracts. All building contracts provide for the standard allocation of risks. The point made in this Latham Requirement is that the risk profile should be capable of being altered, depending on the project. The extent to which the discussion of project-specific issues occurs varies greatly. Providing a mechanism for the recording and mitigation of these risks makes good practical sense whatever the contract be.

A major benefit of adopting pre-commencement contracts is to allow developing risk profiles to be discussed and allocated according to a pre-agreed framework. It is similarly tempting to see a smart contract as being a one-size fits all block of code. However, individual changes in the code and clauses to reflect the site characteristics and constraints are eminently possible. The flexibility and versatility of the smart contract approach can lead to complicated contractual conditions being recorded and set for execution in minutes rather than through hours of expensive lawyer's time recording.

8.1.7 Variations

Taking all reasonable steps to avoid changes to pre-planned works information. But, where variations do occur, they should be priced in advance, with provision for independent adjudication if agreement cannot be reached.

Variation claims account for one of the biggest causes of disputes within the construction industry. Removal of this frequent trigger of disputes requires action on two fronts:

1 Have the client much better briefed on the consequences of a design change beyond the agreed point where the build decision has been taken, and
2 Improve the quality of the data on the design in the construction documents. The less likely there is to be the need for variations. Furthermore, where variations do exist they can be priced more accurately based on the added detail. The parties should be able to build up additional shared information as to underlying costs in order to reduce surprises later on.

Advances in pre-commencement activity and a much greater level of detail on planning should result in variations becoming less common. It is often the client and the funder's fixation with starting on site before they are ready that directly leads to the variations being discussed. The false sense of security that comes from an early start on site is quickly exposed in the mounting list of avoidable variations.

Variations also present a huge challenge for a smart contract. Two of the features of smart contracts are intended to be immutability and it cannot be stopped once commenced. It is not inconceivable that a later prototype smart contract can cope with variations, as this would become an exception-handling procedure possibly for an oracle to order or observe. However, smart contracts are clearly intended in the first instance for a zero-client change of mind scenario based on as much pre-construction, factory building of components and pre-assembly as possible.

8.1.8 Mechanisms for assessing interim payments

Express provision for assessing interim payments… Such arrangements must also be reflected in the related sub-contract documentation. The eventual aim should be to phase out the traditional system of monthly measurement or re-measurement….

Construction contracts have always been measure-and-value arrangements. Simply sending an invoice at the end of a project for services rendered has never been the modus operandi of the industry. The huge risk on the cost of the project and non-availability of sufficient funds have created the norm where the contractor only carries the risk until the first interim payment event. Further, the cost on the client of having to pay the contractor's financing charges is a further dis-incentive to seek longer payment terms.

Advances in technology and, in particular, modelling software involving virtual design and construction have led to improvements in the predictability of earned value and the amount of payments to be made. One benefit of a smart contract approach to self-performing and automated payment functions is that the mile-stones can become inch-stones or even smaller road markers on the way to completion of the project.

8.1.9 Payments

Clearly setting out the period within which interim payments must be made to all participants in the process, failing which they will have an automatic right to compensation, involving payment of interest at a sufficiently heavy rate to deter slow payment.

The importance of releasing cash flow to all parts of the supply chain is a theme which has already been covered. The UK Government mandated payment to tier-3 contractors within 30 days in a welcome recognition of the massive contribution made to the industry by those not in direct contractual arrangements with the client. Late or longer agreed periods of payment to subcontractors is often an indicator of a main contractor being in financial peril. Carillion subcontractors were on 120 days and many are still waiting.

The project bank account has been a great example of a simple yet effective measure to get around the cash flow issues encountered down the supply chain. Rudi Klein has been promoting a campaign to follow the Australian example and to mandate this into law in England and Wales.[7]

8.1.10 Trust funds

Providing for secure trust fund routes of payment.

Trust law is a different animal to standard contract law and is unlikely to be featured in any smart contract arrangement any time soon. Trusts operate for the benefit of the beneficiaries with the trustees taking an altruistic approach to the protection and nurture of trust assets for the accumulation of capital value and interest for later distribution to the beneficiaries. As such, it is at odds with the central purpose of a contract where mutual obligations are exchanged for due consideration flowing both ways.

Nevertheless, the project bank account involves trust law characteristics in its attempts to ring fence the payments from other interested parties or would be creditors. Subcontractors who would have, but for the project bank account, been facing massive bad debt provision on Carillion's collapse were able to enjoy the trust status of money earmarked for them inside the project bank accounts.

8.1.11 Speedy dispute resolution

While taking all possible steps to avoid conflict on site, providing for speedy dispute resolution if any conflict arises, by a pre-determined impartial adjudicator/referee/ expert.

Adjudication has become the speedy dispute-resolution procedure envisaged by Latham. Adjudication still operates as was originally intended although recent cases have led to the odd breach in the dam seeking to hold back the torrent of procedural and other challenges that would wash away its core value – an enforceable decision within 4 weeks or Referral. A discussion of the current position in the United Kingdom on adjudication is contained in section five.

Dispute avoidance, rather than resolution, is a key promise of the potential for smart contracts. There ought to be less disputes based on incompatible contract terms if a smart contract is able to run a self-diagnoses flagging up inconsistent provisions. The disintermediation of the role of the dispute resolver is likely to come into an increasing level of prominence. The Judicial Commission in the United Kingdom has indicated that it will provide whatever form of conflict resolution is desired by the disputants of the future and is willing to be extremely flexible in this connection. The facts in dispute ought to be incontestable by checking the ledger and the consequences that are much more easier to diagnose where they can be checked against a digital twin or as built record generated by overlaying planned and actual in an onsite programme. This is not to say that disputes will be eliminated, people argue over the most trivial of detail and this is frequently personality led and/or down to extraneous causes such as (parent group insolvency than factual disputes. It remains a truism that if there are enough zeros on the end of a sum claimed then it is worth fighting over. In the author's experience, it is often the case that companies pursue claims rather than face the reality of a bad debt or mistake and its potentially catastrophic effect on a balance sheet. Obviously, avoiding such a position being created in the first place is infinitely preferable.

8.1.12 Incentives

Providing for incentives for exceptional performance.

Into the same category of having performed useful service but unlikely to be necessary in a stigmergic future is the incentive contracting approach much beloved by the NEC. Option C is a cost-plus contract that is subject to a pain/gain-share mechanism by reference to an agreed target cost built-up from the activity schedule. The mechanism enables the contractor, and/or the consultant team, to share in the benefits of cost savings but also to bear some of the cost when there are cost overruns. The extent to which the contractor is motivated by the gain share is debatable. In a factual analysis, the contractor's price should not need embellishing though an incentive. In the author's experience, the worker wants the pay packet not the ribbon tying it up.

Particular attention has also to be paid where the original target cost is allowed to run away with itself, as is all too frequently the case with mega projects with spiralling costs. In such circumstances it is at least possible for the incentive schemes to be leveraged in favour of the supply chain.

Key performance indicators (KPIs) are another incentive-based type of performance measurements. KPIs evaluate the success of an organisation or of a particular activity in which it engages. Often, success is simply the repeated, periodic achievement of some levels of operational goal (e.g. zero defects, 10/10 customer satisfaction, etc.), and sometimes success is defined in terms of making progress towards strategic goals. Accordingly, choosing the right KPIs relies upon a good understanding of what is important to the organisation. "What is important" often depends on the Employer's business. These assessments often lead to the identification of potential improvements, so performance indicators are routinely associated with "performance improvement" initiatives. A very common way to choose KPIs is to apply a management framework and to apply targets for improvement.

KPIs on a construction project can include the criteria shown in the example above. Improvement to safety records and defects recorded are two commonly used indicators. A more sustainable approach to construction can also be achieved by measuring power used and waste generated.

KPIs are a good idea and many clients have embraced the approach. However, the proper implementation of the scheme requires resourcing and the partiality of the measurement arrangements also needs to be considered. The notion that continual improvement is relentlessly possible is open to challenge. A point probably exists where the Contractor is performing at a high level and further improvement might be unrealistic. In this situation, the indicators might have served their usefulness and require improving upon themselves.

Another criticism that can be laid at the door of the KPIs is that they are sometimes contradictory. For instance, in higher education around ten years ago, the KPI given prominence was to do with the lecturers not using enough range in their recorded marks and marking in too narrower band. The KPI was on a wider spread of marks, and consequential uplift of the average. This gave way to concerns over grade inflation resulting in the next KPI assessing whether too many top-degree classes were being awarded.

Although KPIs have had a contribution towards embedding partnering, the prediction is that in the techno-enhanced future such statistics on good practice can be reduced in importance as contract parties deliver on their obligations in a timely, mutually dependable and profitable manner.

8.1.13 *Advanced mobilisation payments*

Making provision where appropriate for advance mobilisation payment (if necessary bonded) to contractors and subcontractors, including in respect of offsite prefabricated materials.

The interim payment regime operating in many jurisdictions recognises the importance of cash flow. The supply side is only therefore obliged to carry the cost of the project up until the first interim payment. This situation is not universal however, and advance payments are common in the Middle East and Africa. Where they do occur, there is usually a contractual provision for the crediting of sums advanced against later valuations. The contractor may also have to pay for the privilege of being extended credit by way of an advance payment bond. This operates so as to direct a third-party bondsman to compensate the client in the event that the advance payment never metamorphoses into site work capable of repaying the advance.

Advance payment provisions are likely to be re-examined in the context of smart contracts; particularly in light of the increased focus, there is likely to be on pre-fabrication and factory-built components. Latham's prediction for pre-fabrication was particularly prescient given that he was writing in 1994. Manufacturers, unlike building contractors, are much keener to see payment, at least in part, upfront. This is likely to lead to a fundamental shift in the funding patterns and approach in construction contracts. Ultimately, this will prove to be a good thing. Many of the sources of dispute in the construction industry are essentially squabbles over money during the process.

This chapter now examines other important background provisions not directly contemplated by the Latham report but nevertheless representing noteworthy developments in terms of facilitating further progress towards the digitally enhanced future. What Latham did not foresee when he was writing was the importance of data. This is the single biggest change over the intervening years. The current necessity, as the Hackitt report identifies, is the need for a golden thread of reliable information. Any contract, which facilitates the development and exchange of data at the right time, will deliver against this updated requirement.

8.2 Other important background provisions

This is a selection of contractual vehicles not specifically mentioned by Latham but which have played important roles in bridging the gap from traditional practice towards more progressive practice in the construction industry. There are likely to be indirect benefits flowing from their adoption in terms of technological enhancement.

8.2.1 *Framework agreements*

Clients continuously commissioning construction work are often interested in reducing procurement timescales, learning curves and other risks by using

Figure 8.1 The framework umbrella arrangement

framework agreements. The purpose of a framework is to establish the terms governing contracts to be awarded during a given period on a call-off basis as and when required. The framework contract documents define the scope and possible locations for the works likely to be required and the contract conditions (Figure 8.1).

Framework agreements are therefore essentially the means by which repeat work is awarded to a pre-selected number of contractors who can be said to be protected under this "umbrella" agreement. The client in effect promises all its projects for a fixed period of time to the framework contractors. There are typically 3–5 framework contractors although this can vary depending on the nature of the work. The framework contractors are then awarded the work and "draw down" contracts are entered into. The obligations under the draw-down contract are usually standalone from the over-arching framework agreement. It is possible for the framework agreement to provide for the measurement of KPIs across the performance of one or all of the framework contractors. This essentially leads to competition amongst the framework contractors who can also be encouraged to share best practice between themselves. A framework contractor usually has the expectation rather than the right to be awarded work. The work can be awarded to the framework contractors in rotation or based on previous performance.

Framework contracting has been around for a long time and has provided solid results in terms of savings in time and money when compared with one-off competitive tendering. The advantage to the supplier is that the likelihood of being awarded a project is higher than it would be under an open-procurement process. However, research published by the Civil Engineering Contractors Association (CECA) suggested that frameworks fail to deliver for the contractors and involve unnecessary effort.[8] Framework agreements require enhancement with other features to best deliver on further industry improvement. Fortunately, these enhanced measures were added in 2019 in the form of the Framework Alliance Contract discussed in chapter 9.4.

8.2.2 Open book accounting

Open book accounting is a partnering style arrangement whereby the Contractor provides the Client with a more in-depth than usual appraisal of the financial information relating to the project. Essentially, the Contractor shows the Client what his costs are thereby effectively opening his books to scrutiny. Open book accounting is generally associated with incentivised target cost contracts, management contracts and framework contracts. This arrangement is similar to a cost-reimbursable form of contract. The manner by which the client is given access to the contractor's records and accounting systems involves collecting the data output from the systems. Collaboration and trust are paramount to success. Where costs are shared then there would be a system to pro rata the share incurred on the client's project. The Contractor is reassured that the information will not be used to effect his profit margins as this element is usually ring fenced to encourage openness on the part of the Contractor. The Contractor therefore is awarded a profit percentage (usually around 10 percent) for his management and co-ordination of the subcontractors and suppliers. The Contractor typically must provide details of discounts and savings he has negotiated with his subcontractors and suppliers. The parties need to be clear about whether profit includes such issues as pension contributions and finance charges.

The smart client is able to examine the open book accounts provided by the Contractor and together they can look to make savings in future with the Contractor safe in the knowledge that his profit is protected. There are diverging views of these arrangements including some scepticism about its desirability and usefulness. One view is that the arrangement can be viewed as ultimately self-defeating for the contractor as the Employer wants the benefits of savings that become apparent over time. Ultimately, therefore, even though profit is ring fenced, there will be a smaller sum on which the profit element percentage is applied. Some arrangements avoid this problem by fixing the profit as a sum rather than a percentage of the contract sum as detailed in the open book approach. This, though, marks a departure from a market approach and feels contrived.

A further development on open book accounting is to use the two-stage open book tender approach. The savings and "pairing down" of the contract sum takes place before the client has committed to undertake the construction works. As such, the arms-length nature of the early exchanges about value provide reassurance on the issue of ensuring a competitive approach.

8.2.3 Two-stage open book procurement

This is one of three "new" methods of procurement referred to in the Government Construction Strategy and published in 2014.[9] The other two were cost-led procurement and integrated project insurance. It is debatable whether the three initiatives are properly classed as procurement strategies. The least convincing is cost-led procurement which appears as a continuation of the common sense approach of agreeing to a budget and either sticking to it or improving on it. Integrated project insurance is referred to below and is really better characterised as a post-procurement insurance-based arrangement.

Two-stage open book procurement is the form of contract that has the most convincing claim to being a new approach and retains echoes of two-stage tendering and the partnering approach to the cost-led procurement. The method was trialled successfully on Ministry of Justice BIM-enabled Projects including Cookham Wood using the PPC 2000 form of contract. One difference to existing arrangements is that the Contractor presents open book information during the tender process and the firming up of the price and ring-fenced profit element takes place before the formal full contract is entered into. The supply side organises itself into a consortium brought together for the project, which includes the design team and the specialist subcontractor suppliers. The consortium work together collaboratively to ensure the client is impressed enough to award them the contract. The "can do" attitude instilled at the tender stage is then embedded in the project delivery.

The two-stage bidding process starts with an outline bid and benchmark costs, which are provided to prospective project teams. Following this first stage, the project teams work with the client to develop the proposal and the contract is then awarded at this second stage. The benefit for the client is in working at an early stage with a single, integrated team allowing faster mobilisation.

The benefits of the two-stage open book procurement method include claims as to up to 20 percent cost savings, stakeholder consultation and the potential to appoint small and medium sizes enterprises (SMEs) within the consortium.

A consortium approach is currently restricted to larger projects and is unlikely to translate to medium- and small-sized projects in the short term. The opportunity for a SME company to be selected by other consortium members is also not clear. Two-stage open book has become the Government's preferred approach as recently enhanced by the Framework Alliance Contract.

The consortia approach, with or without a two-stage open book arrangement, would have the greater interest from those planning a smart contract approach.

The tactical grouping together of companies for the delivery of the project, and the hope of repeat work, would confer the infrastructure required for the writing of the necessary automated provisions as well as providing the populous for a distributed ledger system. A private blockchain arrangement would be adequate for the auditing and recording of the execution and performance of the smart contract terms.

8.2.4 Supply chain integration and early contractor involvement

The extremely valuable contribution that subcontractors and suppliers bring to projects in terms of expertise and design input is routinely captured on modern projects. The contrary arrangement, where designers and subcontractors are not consulted ahead of their engagement now appears to be bad practice. These two concepts, at one time viewed as being the exception, are now the rule. All of the standard form of contracts facilitate their inclusion. The term "supply chain" appears to come in and out of common parlance. The image of hierarchical arrangements where the end of the chain is not party to the decisions made higher up is part of the negative connotation as is the slow filtering down of the cash flow to the lower links of the chain. The preferred approach was to view the supply community as a network where the inter-dependence of the disparate elements is recognised. In reality though, it is a chain that hopefully achieves the purpose of linking the project with its necessary components (Figure 8.2).

Early contractor involvement is particularly well suited to large or complex project and can be facilitated by the use of a pre-construction service agreement that enables the client to employ a contractor before the main construction contract has commenced. A perceived disadvantage of engaging with the contractor too early is that they are unable to give an accurate price. Whilst the risk of price rises can be off-set by other measures, such as two-stage open book procedures, the reality is often that the embedded contractor has a significant advantage the longer their involvement lasts and the interest of other contactor's is lost.

In supply chains where there are particular vulnerabilities, enlightened clients recognise the value in protecting the weak links in that chain. In certain Ministry of Defence projects, there existed a "fragility index" in terms of flagging up when a vital component supplier was experiencing financial issues. This would then trigger a conversation and a discussion of support required to see the supplier through a rough patch or a short-term cash flow crisis. This is a very progressive approach to take and has much to commend it. However, the limits of co-operation would need to be defined at the risk of the client becoming too complicit in the business' survival and management.

Another issue to consider around supply chains is that when they are extended, they can also result in some practices that are at odds with sustainable practice. Responsible sourcing of components including having regard to how many miles a product has to travel would be part of a provenance-based approach delivered as a part of the traceability criteria of modern procurement methods.

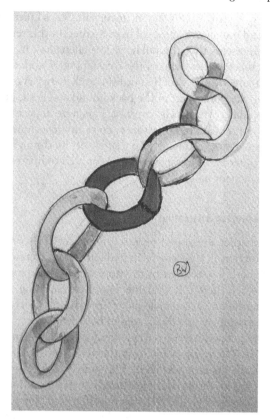

Figure 8.2 Supply chain

8.2.5 The pipeline

The National Infrastructure and Construction Pipeline became the single course of the United Kingdom's planned infrastructure investment to 2021 and beyond. The projection of infrastructure investment over the period 2019–2029 was £600 billion and as at 2020 there are currently 700 projects in the pipeline. Importantly, the pipeline covers projects funded in the public (local and central government), private and mixed sectors. Publishing the pipeline enables visibility knowledge and understanding of where infrastructure investment if being made and by whom. The express aim was to boost market confidence and help the sector with business planning. The government also sought to commit to a low-carbon economy and to promote innovation.

The pipeline involves a complete centrally accessible data set with project particulars. All government departments maintain the pipeline and identify value for money criteria relevant to the project to be converted to standards and specifications for suppliers. The issue about whether SMEs are genuinely able to

compete effectively for this work remains uncertain. Steps taken to simplify tender information and framework approval have assisted in this regard.

The pipeline represents that the public sector client has, in the popular parlance, "got its act together" in terms of the throughput of work and that part the
private sector can directly affect such as utilities and energy. Again, this achievement is something that forms part of the background and is a legitimate expectation for a mature industry in the 21st century. The more impressive aspect of this
is how the public sector client has now sort to co-ordinate its frameworks under its
different ministries and departments into a single entity through the FAC-1 that
is discussed in chapter 9.3. The opportunities for a co-ordinated smart contract
approach are also evident here.

8.2.6 Integrated project insurance

The subject of whether integrated project insurance (IPI) or similar move to
blame-free construction is a necessary ingredient in the technological future is
a vexed one. The author sees merit in this approach and can foresee a construction industry in which there is zero blame. However, retaining accountability in
the short-to-medium term is extremely likely. Integrated project insurance is not
therefore characterised here as a direct step in-between.

Integrated project insurance was a new kind of procurement trialled by the
UK government in the years following its unveiling in 2011. There are very
few reported projects where it was used. The most well-known example in the
United Kingdom was at Terminal 5 Heathrow construction. More recently, the
Built Environment School in Dudley was another exemplar project using this
approach. The idea is simple – rather than oblige the participants to carry their
own multiple insurance policies, there would be only one policy. A helpful analogy here is that of a car park. At present, all the, say, 400 cars in the car park
have their own insurance. If a collision occurs between two drivers, then there is
a fault-based discussion and insurers are called. Most of the time, the outcome of
that discussion can be a "knock for knock" basis where the companies agree to
owe each other one for next time and take care of their own insured. This is not
very far removed from the IPI approach but without the need for 400 policies. All
of the policies represent an overhead for each individual driver which could be
replaced with the single over-arching policy.

The blame-free construction has its critics in terms of the down side of removing accountability which of itself is part and parcel of professionalism. Other
prefer a more nuanced approach to better practice based on collaboration rather
than a carte blanche to operate as one wishes.

Nevertheless, if autonomous vehicles and machines capable of implementing
their own machine learning are involved in a construction project then it is difficult to see on what other basis law can facilitate their interactions with regular participants other than through an integrated insurance approach with strict liability.

The idea to replace multiple separate policies of insurance held by the project
stakeholders with a single policy is at the core of integrated project insurance.

Current provision often have overlap and duplication resulting in waste. However, contractors and professionals alike are attached to their policies as the final defence against what could be catastrophic consequences for their businesses if they were not so protected. Having sufficient trust in a policy procured by a third party requires a leap of faith on the part of the supply chain and professionals appointed. One outcome might be further duplication as the IPI and a separate policy are maintained. One precaution against this would be to name all of the parties as co-insured. There is unlikely to be much appetite for this amongst insurance companies at present due to wariness over extending the cover for too many organisations.

Various case studies have recorded successful use of the integrated project insurance most prominent of which is the Dudley Built Environment College. It is worth examining the published material on the College in order to gain a holistic view on how the latest best practices can operate together.

8.2.7 Dudley College

This new educational establishment has a project value of £11.69 million opened in 2018 and featured an alliance arrangement with constituent members being specialist engineers, architect, structural engineer and the contractor.[10] Key suppliers including steelwork, lifts, metal decking, ceilings and partitions were also included in the alliance. The IPI brokers were involved throughout the project in order to facilitate and provide assurance on the risks being covered. The philosophy of the project was that the alliance viewed itself as a "virtual company." All alliance members waived their right to make claims against each other save to the extent that there was "wilful default." The looseness of this important legal term would appear to have been a risk for the parties.

The project also had a project bank account and BIM. Further, cost, time and carbon savings were in line with the Government Industrial Strategy: Construction 2025. The whole team was appointed at inception. It is in the integration of these measures that the most promising signs exist for planning a future direction for the construction industry.

However, with the alleged cost of the integrated policy at around 4 percent of the contract value it was an expensive exercise. Further, the potential financial exposure of the insurer was capped with the client picking up any exposure passed the limit. Below the upper limit, a series of pain-and-gain-share arrangements were implemented subject to individually negotiated limits. This risk may be unacceptable to other clients and contractors.

On the positive side, the alliance members were able to take part in advanced BIM work free of any inhibitions about liability. Reportedly the SMEs in particular felt liberated and made telling contributions way and above what they would normally proffer. This led to a solutions-based approach rather than finger pointing and covering any perceived exposures. The project recorded impressive percentage cost savings on its initial budget.

An important aspect of existing insurance arrangements is the right of subrogation whereby the insurer can avail itself of any rights against third parties that the insured may have to recuperate any pay-out made. This right is seen as being valuable to the insurers but was waived on the Dudley College project. The project also contained a fitness for purpose warranty rather than the usual reasonable skill-and-care obligation. This is an important development towards smart contracts that must surely use the higher standard. However, this is still seen as a step too far for many insurers who remain extremely nervous around the higher standard given the increased risk to them of settling claims in much wider circumstances.

The project itself stands as an example of what can be done with a committed client able to deliver hands-on top-heavy management in the early stages. Realistically, its suitability is for a fairly narrow band of projects. The scrutiny required by the insurance brokers' agent is also an additional overhead comprising the vetting of the design and cost risks ahead of the project commencement. The vetting was, no doubt, considered necessary as the insurers might make an assumption that with the waiving of subrogation might also come a waiver of skill and care. The blame-free environment that might result may see professionals adopt a less professional attitude. Perhaps, the reverse is true as was the experience of the SMEs going further in the enlightened environment they found themselves in. Ultimately, smart contracts and automated construction should render the distinction meaningless. The granularity and planning involved in a meticulously delivered smart contract should ensure either a complete absence of mistakes or a reliable data record to the exact consequence and root cause of the mistake, which are themselves quickly rectified. For now, professionalism is a cornerstone within the built environment and something that will be very hard to relinquish.

It has been suggested that a similar "blame free" or "blame light" approach is required in order for hub-and-spoke models including BIM and smart contracts to work properly. However, this can be viewed as an over-simplification of the role of insurance. Insuring a risk is only one of Abrahamson's original strategies for dealing with risk. Risk management primarily involves avoiding or reducing risk through agreed actions. This is supported by insurance but should not be subservient to it. Dudley College approach is not the only way to avoid claims and return costs and time savings on projects.

It is a fact that claims and disputes mar the construction industry. Any attempt to reduce this is to be welcomed. However, to remove accountability and professionalism from the equation without due consideration of what will replace it appears a little short of providing the complete solution intended. It is interesting therefore, in the discussions around BIM in chapter 10, that liability for mistakes remains part of the approach being taken at present rather than a blame-free scenario.

One way which has been found to proceed whilst maintain some "blame" in the process has been trialled in the United States where "wrap" insurance is found. The Employer pays the premium on the integrated insurance policy. Each participant pays a contribution by way of a reduction in their contract price. The percentage contribution of the party is not fixed and subject to re-allocation in the event of a claim. In essence, any negligent party would be retrospectively allocated more of the premium paid. Further, the participant's reputation will be damaged by any allocation of the premium leading to all possible steps being taken to avoid being "at fault." The potential penalties here are manageable and insurable rather than amounting to business-threatening liabilities as per the risks on a normal project.

Insurers in the United Kingdom have been very slow to offer integrated project insurance despite government encouragement. The non-availability of these products is of major concern to the establishment of this procurement route.

Chapter 9 moves on to consider the steps that have been identified as definitely contributing to the digital future. Not only do these steps meet the criteria for justification, but they also focus on shifting people's attitudes through an interactive process whereby exposure to these initiatives cannot help but convince those involved of the truth of the statement that performing the tasks any other way is to behave nonsensically.

Notes

1. Lord, W. (2008) *Embracing a Modern Contract – Progression since Latham?*, RICS Conference Proceedings.
2. Mosey, D. & Jackson, S. (2020) *Good Faith and Relational Contracting – Do Enterprise Contracts Offer a Way through the Woods?*, SCL paper D 228.
3. *Wilmot Dixon Housing Limited v Newlon Housing Trust* [2013] EWHC 798 (TCC).
4. *TSG Building Services Plc v South Anglia Housing Limited* [2013] EWHC 1151.
5. *Interfoto Picture Library Limited v Stiletto Visual Programs Limited* [1989] QB 433.
6. Fischer et al. (2017) *Integrating Project Delivery*. Wiley, USA.
7. Klein, R. (2019) *What will the Project Bank Account Bill Do?*, Building Magazine 6 February.
8. https://www.ceca.co.uk/ceca-time-for-government-to-get-a-grip-of-frameworks/.
9. Cabinet Office (2014) *New Models of Construction Procurement*. HMSO.
10. https://constructingexcellence.org.uk/wp-content/uploads/2015/12/Trial-Projects-Dudley-College-Advance-II-Case-Study_Final.pdf.

9 The steps in-between

The triple jump is a discipline in athletics that requires the combination of three component parts – the hop, skip and jump (Figure 9.1). This analogy works well for the type of co-ordination necessary for the fulfilment of the desire to move from the AS IS position to the TO BE position in relation to smart contracts and digital construction law.

"*Smart contracts are not business as usual, but they are not the end of contract law as we know it, either.*"[1] The normal way in which developments are made is through the refurbishment of existing concepts and the progression of current ideas in the wake of a social and technological change. This section is a review of the current best-practice ideas and a discussion on how they could assist in the development of smart contracts and vice versa how may smart contracts themselves aid these concepts to achieve their intended or extended outcomes.

9.1 Government support

In 2013, the Government recognised the importance of the construction industry to the UK economy including as it does 280,000 businesses and 3.2 million jobs (9 percent of total employment). The report set ambitious targets for 2025 to cut project costs by 30 percent and the time they take for completion by 50 percent. Recognition was given to the fact that this would only be possible if the industry was prepared to digitally transform itself.

The 2018 Construction Sector Deal took the agenda further in prioritising the development and commercialisation of digital and off-site manufacturing technologies aimed at producing safer, energy-efficient buildings that perform better through their life cycle.

The obstacle for both reports is that digitisation is not about science but about people. The industry continues to struggle to work together well and to embed new initiatives into its practices and regulatory frameworks. It is for the clients to articulate with better contracts, guidance and standards to enable so suppliers to be clearer about what is expected.

Significant parts of the industry seem to take a reactive approach to digital upskilling for clients to demand it rather than taking a more proactive and visionary role. This is what was experienced on the Dudley college project.

Figure 9.1 A hop, skip and jump

Notwithstanding the blame-light environment, the parties needed a considerable coaxing before embracing the new philosophy.

Construction's reactionary nature means that it has not gone through any form of a productivity or innovation phase and still represents low-hanging fruit for the application of technology and process improvement. According to Mark Farmer,[2] *"Digital automation and augmentation of the industry's current proliferation of labour-intensive, low efficiency processes are now in the cross-hairs of tech start-ups, investors and entrepreneurs who rightly smell an opportunity."*

9.2 Digital upskilling

The meaning of digital skills can be broadly defined to mean any skills relating to being digitally literate. All of us are at a different point on a spectrum of ability – for some, this might mean checking your social media on your smart phone whilst, for others, it might mean learning to code a website or even a smart contract. The construction industry as a whole has a poor level of digital skill and improving on this position is an important step in-between the AS IS and TO BE position. Part of the issue is around whether the clients are being presented with a compelling enough business case to adopt the new technologies. Clients wish to see value added through the technology and the signs are that this will come with time. Data gathering will only increase in importance and being able to demonstrate issues like embedded carbon and building performance will trigger tax incentives and increasingly preferential status on the early adopters. This is demonstrable with those seeking the standard set by the Building Research Establishment Environmental Assessment Method (BREEAM).[3]

For many, digital upskilling is an iterative process and the best and most successful technologies are those that are easy to use. For example, if I am familiar with one online meeting programme then I should be able to use another without too much trouble. The new programme will have "upskilled" me and potentially expanded my grasp of its functions such as share screen that the other programme might not have had.

Recent digital skills development in the construction sector should take this iterative approach and embed it in our relationships with technology. An interesting take on this comes from Trimble,[4] who envision all procedures and processes around construction being as easy to use as social media and our personal time management such as online banking and meter reading. When technology is as much second nature as uploading a photo onto a group chat and moving money as easy as paying a friend through paypal, then construction will have upskilled to the required extent. The same organisation presents the following areas of digital skill:

- Digital and Manufacturing Technologies – For example, a building information modelling (BIM) file of a steel structure for a project is exportable to Excel or capable of being sent directly to fabrication machinery. The same data then can be used for estimation, erection sequencing and ordering.

- Digital Design – The opportunity to examine and use the digital design for other purposes is currently demonstrated by programmes that can embed net zero carbon targets, trace emissions and noise generation tied into the BIM file.
- Digital technologies

A simple example of where digital technologies can assist is in the use of cloud-based site diaries. Weather, personnel and equipment records would be created and stored on the cloud. In a later iteration of this technology, there could be a link to smart contracts as the agreed site diary becomes the data set from which the transactions are performed. Other more advanced technologies include laser scanning, cloud-based near-miss reporting and snagging apps which can be ticked off once complete.

Augmented and virtual reality. Augmented reality uses digital information which is overlaid on the real world through design models (BIM). Virtual reality is immersive and allows the user to put on a head set and walk through the building before it is built. This can be extremely useful for the layout of such things as operating theatres in hospitals where the surgeon can pre-agree the layout of the facility. The ownership of designs stored in virtual realities remains an issue for those concerned about the loss of their intellectual property rights. Increasingly, this immersive technology uses haptics or kinaesthetic communication. This refers to any technology that can create an experience of touch by applying forces, vibrations or motions to the user. This technology is already well established amongst gaming platforms. Practical applications of augmented reality in construction include Trimble Site Vision that allows the user to access a projection of the BIM platform against the real world through the use of their smart phone. Similarly, the location of existing utilities can be seen against the real world through the use of a tablet or similar.

3D printing is a reality is many industries and construction is no exemption. The ability to make a building component on site for immediate installation is now possible. Neither do 3D printers use solely plastic as the printing medium. Airbus industries apparently print in titanium and carbon fibre. Digital fabrication has been used to make a car and even whole buildings are capable of 3D printing. The challenge for the supply community in protecting its products and the warranties that would accompany any 3D printed product would need to be addressed. 4D printing is similar to 3D except that the printed item only assumes the desired shape at the point at which it is to be used. A cube structure may therefore be 3D printed and transported flat pack only too spring into a cube when unpacked.

Robotic construction is similarly no longer in the realm of science fiction. Driverless trucks operate in remote mining operations using machine-control technology where there is little if any human involvement in an entire running plant. It is straightforward to envisage that BIM data may evolve to the extent where robots need merely to access the model in order to receive their instruction on what to build and where. The legal challenges around these issues would mean that negligence-type liability would need to be replaced with the higher fitness for purpose duty. Companies offer self-driving construction machinery to perform repetitive tasks more efficiently than their human counterparts, such as pouring concrete, bricklaying, welding and demolition. Excavation and preparation work is being performed by autonomous or semi-autonomous bulldozers, which can prepare a job site with the help of a human programmer to exact specifications. This frees up human workers for the construction work itself and reduces the overall time required to complete the project. Project managers can also track job site work in real time. This is achieved by using facial recognition, onsite cameras, and similar technologies to assess worker productivity and conformance to procedures.

Augmented workers using exo-skeletons for enhanced performance are the subject of interest in the construction sector. This would allow the productivity rates of a worker to improve and lead to quicker and cheaper construction. The liability for any injuries caused by the technology would require insurance cover.

Wearable Technology. The workers' kit can plot stress and heart rate and lead to intervention where rates reached dangerous levels. Location, fatigue levels, humidity and temperature can also be monitored. One positive result which is achieved when the results are shared with the workers is an improvement in fitness. Wearable technology such as sensors and location trackers assist with site management and real-time updates. Legal issues arising out of this technology include accessing personal data that is protected by legislation in form of the Data Protection Act 1998.

Drones can now be used to measure quantities in a labour-saving breakthrough for the industry. Groundwork measurement of extracted material or façade-condition reporting may currently take weeks to measure using conventional methods whereas a drone can perform this task in a matter of minutes. Similarly, ground-penetrating radar can remove the risk of unforeseen ground conditions.

Big Data/Internet of Things. The technology exists to provide the information needed for the blockchain to be gathered by its multiple reference points. Massive amounts of data are created every day. Every project becomes a potential data source for artificial intelligence (AI). Data generated from images captured from mobile devices, drone videos, security sensors, BIM and others forms a pool of information. This presents an opportunity for construction industry professionals and customers to analyse and benefit from the insights generated from the data with the help of AI and machine learning systems. The embedding of censors in devices is already in wide usage and is set to pass 25 billion by 2020. Project managers are able to monitor deliveries of materials by tracking the location information embedded in the goods in transit. Project managers maintain a dashboard of building performance based on information being sent from installed plant and other products. It appears logical that the censors should report back to the federated BIM model where planned completion is over-written with actual performance on the project.

The work of Digital Catapult is promoting distributed ledger technology in construction in the United Kingdom. Their mission to actually engage real uses for the technology is commendable. Under the banner "Advanced Digital Technologies," four fields of practice are identified which are:

- future networks, using 5G and the internet of things (IOT),
- Immersive, virtual reality, and augmented and mixed reality and haptics
- AI and machine learning
- Distributed systems, DLT and blockchain

A specific construction project known as the Weather Ledger has already been mentioned whereby the promotion of real-world use of smart contracts in construction is sought. The idea is to automate the weather compensations clauses and conditions in standard construction contracts. The clauses are analysed and converted into smart contract provisions that are visible to all stakeholders. The details of the sources of the weather data, the regularity of updates, the parties to be informed and an external arbiter are all set out. The weather example has been chosen partially because there are no data protections issues and no sensitive data and the issue is simple – the weather is either more compensable or it is not.

The mechanics of the project are that the data is picked up from the trusted online weather service and fed into an application programming interface (API) or software intermediary. The API then converts the event into a trigger event for a smart contract that is then verified on the blockchain leading to instant awareness and immediate compensation through the automated payment mechanism. The research into the project and its key deliverables are ongoing as at 2020.

A more tangible Digital Catapult project involves the Bloodhound project (Figure 9.2) originally developed in part at UWE Bristol.

In its latest incarnation, the Bloodhound is in 2020 preparing for their attempt at a new land speed record of over 500 mph. IOT sensors will be located at regular

Figure 9.2 The Bloodhound car

intervals along the 12-mile race track to aid the Bloodhound team in understanding how weather patterns are likely to impact the challenge outcome.[5] The sensors are battery-powered devices which can run continuously for a year and operate with very little infrastructure. The technical specifications of the set-up have many direct correlations with the systems that can be used in construction projects.

One criticism of the distributed-ledger-technology approach and IOT sensors is said to be that it represents too much of a drain on processing power, requiring, as it does all the nodes to approve the transactions. DAG stands for "Directed Acyclic Graph" which provides for two randomly selected nodes in the programme to verify the transaction. Supporters of this approach identify that the security remains as good here whilst drastically cutting down on the power usage.

9.3 Smart technologies – the here and now

This section considers the here and now of what initiatives are actually in use and employing the technologies discussed. Whilst this does not amount to wide scale adoption, it does represent steps in-between.

One of the important developments in machine learning relates to events already mentioned. The computer victory in chess in 1997 and in Go in 2016 was a chance to review the progress between the two events and what this represented. The difference was that in the computer that won the chess matches, it did so with machine power systematically evaluating 200 million possible moves on the chessboard per second and winning with brute number-crunching force. There was no human creativity or intuition. The difference was that Alpha Go used algorithms to find its moves based on knowledge previously "learned" by machine learning, specifically by an artificial neural network both from human and computer play. This machine learning aspect is of great interest to the built environment and has numerous applications already.

AI is an aggregative term for describing when a machine mimics human cognitive functions, like problem-solving, pattern recognition and learning. Machine learning is a subset of AI. Machine learning is a field of AI that uses statistical techniques to give computer systems the ability to "learn" from data, without being explicitly programmed. A machine becomes better at understanding and providing insights as it is exposed to more data. An example of this could relate to the maintenance of a building, the replacement of components and the interactions of users of the building, and how best to accommodate their needs and patterns.

The potential applications of machine learning and AI in construction are vast. Requests for information, open issues and change orders are standard in the industry. Machine learning is like a smart assistant that can scrutinize these mountains of data. It then alerts project managers about the critical things that need their attention. Several applications already use AI in this way. Its benefits range from mundane filtering of spam emails to advanced safety monitoring.

Most mega projects exceed their budgets despite employing experienced project teams and setting out with the best of intentions. Artificial Neural Networks are used on projects to predict cost overruns based on factors such as project size, contract type and the competence level of project managers. There is great support for this movement internationally including comparisons of school building construction costs in South East Asia and apartment projects in Vietnam and building projects in the Philippines. Historical data, such as planned start and end dates, is used by predictive models to envision realistic timelines for future projects. AI helps staff remotely access real-life training material, which helps them enhance their skills and knowledge quickly. This reduces the time taken to on board new resources onto projects. As a result, project delivery is expedited.

9.4 Project Bank Accounts (PBAs)

PBA can be incorporated into any of the other contract families whereby the existing payment arrangements are effectively bypassed and put through the PBA.

The main advantages of the PBA are felt by the supply side, most notably by the subcontractors and suppliers. The interim (or stage) payment, once certified by the overseer, is transferred into the PBA. The bank account only pays out the sum to the main Contractor and the sub-contractor when the former and latter agree upon the amounts properly payable to them. Essentially, the main Contractor does not access his payment until he has also agreed how much the designated subcontractors (not all of whom will be PBA holders) will also receive. It is by this strategy that the government seeks to fulfil its commitment to pay down to tier-3 contractors within 30 days. The deal for the main Contractor is essentially if they do not agree payments both upstream and downstream then they are kept out of their own money. The arrangement has a good deal to commend it in terms of taking out hidden costs and pressures from the industry bearing in mind Lord Denning's prophetic words about cash flow being the life blood of industry.

Guidance for setting up and operating PBAs was first published in September 2007 by the Office of Government Commerce. The guidance recommend that central government clients used PBAs. PBAs prevent the situation, as per the Carillion episode, where the main contractors become the unofficial banks for subcontractor money. The money is paid from one central bank account, which removes the need of having the money cascading down through the layers of the supply chain with the consequent risks that the money will be paid late or not at all. All project participants are paid simultaneously. This helpful table (Table 9.1) from Rudi Klein[6] makes the point:

Table 9.1 Payment bank accounts

Traditional payments	PBA payments
Client to Tier-1 Contractor 30 days Tier-1 Contractor to Tier-2 Contractor 30 days Tier-2 Contractor to Tier-3 Contractor 30 days TOTAL 90 days	Client to PBA 30 days Tier-1, 2 & 3 contractors 5 days* TOTAL 35 days • Assuming the maximum length of time for monies to be in the PBA is 5 days

The PBA provides a safe haven for the monies intended for the supply chain. It is not the means by which contractual entitlement is established as this is the function of the relevant contractual provisions. In this sense, the PBA sits on top of the payment provisions of the underlying contracts. The tier-1 contractor usually sets up the account and is either the sole account holder or opens the account in the dual names of the contractor and the client. The money in the account has trust status and the supply chain are the envisaged beneficiaries.

Highways England have been a pioneer in the use of PBA and have paid for £20 billion worth of highways work in this manner. The Environment Agency are also frequent users as are various Australian states including Western Australia, Northern Territories and Queensland. In Queensland, the legislature have taken the bold step to mandate the use of PBA for the private as well as the public sector. It remains to be seen whether the UK Government are feeling similarly emboldened to make a similar law for our private sector. The Framework Alliance Contract (FAC) estimated £30 billion use from the public purse mandates their use.

The experience across the United Kingdom of the use of PBAs has been positive. Quicker and more regular payments are made and a measure of protection against upstream insolvency is delivered. Their ability to promote collaboration is also self-evident given that all project participants are paid from the same account on which they have transparency.

The PBA removes the temptation for the main contractor to seek additional profit by covertly replacing approved supply chain members with cheaper alternatives. An incidental effect is to make the client aware of any changes in those members paid by the PBA.

PBAs typify what can be achieved by a solution that is simple yet transformative and completely compatible with other developments in intuitive response to the problems of the industry. The same compelling approach and logic supports the adoption of distributed ledger technology and, ultimately, smart contracts which, resonate with the key themes of transparency, traceability and collaboration.

9.5 New Engineering Contract (NEC4)

The underlying philosophy of the NEC suite of contracts is their commitment to proactive risk management, open communication and the willingness to work together to solve problems. The extent of the penetration of the NEC into the standard form market continues to divide opinion, partially because of the changing metrics of the leading contracts in use survey which reports on a decline in NEC usage as the most often used contract.[7] The statistic here sees the Joint Contract Tribunal (JCT) names as the most often used by 62 percent and NEC 14 percent. This statistic masks the fact that NEC is likely to be much more prevalent on large projects, which is not picked up on this metric. More positively for NEC supporters, it was used by 39 percent of respondents at some stage in the past 12 months.

The NEC4 suite of contracts, published in June 2017, replaced the NEC3 suite and introduced new contracts including a professional service subcontract and a term service subcontract. Both of these new contracts address integration of the supply chain including the designers. In another development, there is also a new Design Build Operate (DBO) contract that addressed construction and operational requirements and recognises the importance of maintenance (see section 7.3). In addition, the NEC4 introduces the Alliance Contract (ALC) with its multi-party alliance featuring an integrated risk and reward model.

The NEC4 follows the same structure and principles as its earlier editions but builds in new options for initiatives including BIM, value engineering proposals and dispute review boards.

The NEC4 reflects user feedback and industry developments and continues to stimulate good project management. The contract is endorsed by the Cabinet office and its contribution acknowledge to behavioural change and embrace of the digital future. There is considerable alignment with the Government Construction Strategy, BIM and Soft Landings. The improvements embodied in the NEC4 were well received by the construction industry. However, it is likely that multiple clients will still insert numerous Z clauses to strengthen their position under the guise of adding clarity to the parties' potential liabilities under the contract.

The NEC has provided impetus to the movement for change and successfully taken on the quasi-monopoly enjoyed in the United Kingdom by the JCT standard forms. This healthy competition and difference in style and approach have led to contract advancement for the benefit of the industry as a whole. Cross-pollination of ideas has occurred and, as a result, construction professionals have choice both as to which contract they use and the options of how to approach issues within the contracts themselves.

A further contractual innovation was introduced in 2019 in the form of the FAC, which amounts to a further challenge to the status quo enjoyed by JCT and NEC forms.

9.6 Framework Alliance Contract (FAC-1)

Collaborative practice is taken further towards its goals by the FAC-1 suite of contracts. The new FAC-1 contract contains all the ingredients of partnering and is an important development which, if used properly, can massively benefit the construction industry and deliver on the agenda for change. The throughput of work is the vital ingredient to ensure commitment to the stated goals of the framework. This is the embodiment of the integration envisaged.

The FAC-1 was written following consultation with 120 engineering and construction clients, consultants and contractors. It is described as "a thoroughly modern contract."[8]

However, the inclusion of the FAC-1 in the steps towards the smart contract is not based solely on its collaborative credentials. Its status as groundbreaking relies more on the enabling it does towards integration of the processes and relationships around the data and technological advances it seeks to facilitate. Further, if the stack approach to smart contracts is to prove the way forward then the schedule approach of the FAC-1 lends itself to the partial development of some automated provisions "nesting" inside a more traditional approach to contracting.

FAC is an umbrella arrangement to enable, connect and enhance multiple project appointments. There is an alliance manager to integrate the alliance, monitor performance and support joint activities within the core group role who. Crucially, it seeks to create and capture enough data to award, compare and integrate project contracts, and to improve on the assumptions made when the FAC was entered into.

The following features of the FAC-1 are illustrative (Table 9.2) and a commentary is provided alongside each point.

The FAC is exactly that – a framework drawing together all the best practice initiatives that pass the acid test of being immediately and uncontestably useful to the construction industry. The "joined up thinking" is its main strength and

Table 9.2 Framework alliance contract features

A multi-party contract	This overcomes the many limitations of the linear approach.
Schedule 1 agreed objectives, success measures, targets and incentives	This schedule is given due prominence with a focus of minds on the key deliverables.
Schedule 2 is a timetable for seeking improved value including consents and approvals required before subsequent actions	Pre-commencement activity of this nature is essential to avoid the problems of "just make a start."
Schedule 3 is a risk register	The creation of a bespoke set of risks and how they are managed is good practice and something that a smart contract programmer could look to personalise on a technology enabled programme.
Schedule 4 is the direct award procedures and/or competitive award procedure	Giving options as to how the work is tendered or not if required gives flexibility and moves away from an arrangement that is too "cosy."
Schedule 5 is the template of project documents	The use of templates is likely to become of increasing importance in setting the parameters for automated terms.
Schedule 6 represents legal requirements and special terms for addition and amendment to the contract terms	Allowing for characteristics of the project whilst providing for set amendments limits the scope for rogue contractual drafting. In terms of smart contract drafting, it will be similarly re-assuring to pre-agree the range of terms that could be featured.
Schedule 7 is the framework brief with all requirements including BIM, supply chain engagement	A core value of the FAC-1 is the integration of BIM. Many other contracts see it as an add-on rather than the portal function it can provide.
Framework prices and framework proposals (remaining confidential between alliance manager and the party submitting the prices)	Maintaining competitive advantage for the suppliers in not sharing their cost breakdowns will be re-assuring provided the same is not breached.

enabler. The challenge remains to transpose the approach into the private sector. It is precisely this approach of having everything in place to promote collaboration that will be required in the smart contract journey. The incremental accumulation of contributing and complimentary technologies and procedures should also lead inexorably towards progress.

9.7 Pre-construction service agreements

Letters of intent regularly feature in the list of things likely to result in disputes in the construction sector. The letter of intent is an all too common reminder of the shortcomings of many project teams to be ready in time to build within their timeframes. Act in haste and repent at leisure is a popular aphorism. This is embodied in the letter of intent approach, which is tantamount to admitting that the project involved poor preparation or at least failed to tackle those issues preventing a signed contract being in place. The legal background to letters of intent is described in the companion book *Construction Law: From Beginner to Practitioner*.

In a situation where a letter of intent is used, it usually follows that the other indicators of best practice are also absent. And yet, there remains a category of project where a letter of intent type solution might be needed, such as where a planning condition requires discharge or confirmation or a missing element of the design requires verification. However, in these situations and in more general practice the use of pre-construction services agreements (PCSA) ought to the norm as the symbol of good practice and instigator of the data management and information flow that will last before, during and after the build as required.

The benefits of using a PCSA include setting a timetable for the pre-construction-phase activities and defining the relationships with consultants, for example, interfaces with designers for the purpose of agreeing value engineering and buildability. Further, the PCSA can place limits on contractor's authority to incur costs (one of the main reasons for using a letter of intent) and set out the procedure to agree the finalisations of detailed designs and the prices of each element of the works. There can be a seamless lead into the timing and procedures for finalising the construction phase contract including its start and finish dates.

The standard form contract writing bodies supply precedents for PCSA use. For example, The JCT version can be used for two-stage tendering involving a fixed fee proposal for pre-construction services, a programme and method statement. The offer comprises overheads, profits and preliminaries rather than a lump sum price. Once selected, the client and preferred contractor co-operate as the latter tenders the works packages for fixed price lump sums. The client then decides whether to accept and enter a lump sum contract for the whole works.

The PCSA contains standard obligations in relation to skill and care, prohibited materials, insurance, confidentiality and co-operation amongst the project team. Their use is also thought to be able to contain any opportunistic attempts by the contractor to subsequently inflate costs or exploit the position under a letter of intent.

It is hard to imagine a scenario where there would not be PCSA type agreements where smart contract adoption is being contemplated alongside framework arrangements and the necessary infrastructure for a long-term relationship. The client using a PCSA is genuinely interested in the contractor's input on buildability, programming, project management and the division of the works into appropriate sub-contract packages. It is much less likely that a client will get the best out of a sophisticated contractor without using the focus of a PCSA.

9.8 Off-site manufacture for construction and pre-fabrication

The construction process has long been associated with the assembly of materials on a site by skilled operatives. Pre-fabricated solutions have provided a viable alternative to this approach for hundreds of years and examples range from Roman forts to post World War II housing. The common feature is the building of elements of the project off-site in a factory or workshop and fitted together on-site which cuts down on cost, time and the labour needed to create a structure. Housing units have long been considered an obvious target for off-site construction. The ambitious targets set for house buying are certainly driving the present day use of off-site manufacture

Off-site manufacture (OSM) is an example of a modern method of construction and refers to the completion of elements or components at different locations whether this be permanent manufacturing facilities or "flying factories" which is the name given to temporary facilities that operate for the duration of a project and then "fly" to a new location to service the next project. Taken to its extreme, this can result in 3D printed homes with several prototypes available and the real prospect of a break into the mainstream in the near future.

Designers, contractors and clients continue to explore how pre-fabrication can solve the pressing problems of the built environment in the 21st century. The arrival of BIM, 3D printing and automation leads to further possibilities in terms of better safety, reduced waste, higher quality and reduced downtime. Off-site construction is particularly suited to high-volume, repetitive components or products that require factory conditions to achieve the desired level of quality.

Mark Farmer's 2017 report "Modernise or die" adopted the term "pre-manufacture" as "…a generic term to embrace all processes which reduce the level of on-site labour intensity and delivery risk," and suggested that this could range from "…component level standardisation and lean processes through to completely pre-finished volumetric solutions."[9]

Pre-fabrication has its downsides in terms of high initial set-up costs and the challenge to maintain a sufficiently consistent pipeline of demand to suit assembly line production methods. Transport costs and site access also require consideration. Further, lingering concerns remain regarding the label "prefabricated" as a result of poor-quality mass-produced housing that was pre-fabricated following the Second World War.

Where off site factories are used, they are increasingly relying on off-site factories staffed by autonomous robots that piece together components of a building, which are then pieced together by human workers on-site. Structures including walls can be completed assembly-line style by autonomous machinery more efficiently than their human counterparts, leaving human workers to finish the detail work like plumbing and, electrical systems when the structure is fitted together.

The relationship between off-site manufacture for construction and smart contracts reveals some interesting points and synergies. One often-cited problem with smart contracts are that they cannot cope with the extraneous factors encountered on a building site that is a unique uncontrolled environment on every occasion. This limitation is not present in a factory setting and this might provide the test bed for smart contract assembly of the component parts in a controlled setting.

Carrying out construction activity in a factory setting is to change the nature of construction contracts to a considerable degree. Payment terms would need to reflect upfront delivery and assembly charges and this might lead to an uptake in vesting certificates and advance payment bonds. Warranties and guarantees would likely become much more commonplace to cover the maintenance and costs in use risk. The insolvency risk of a manufacturer of buildings would require security checks and parent company arrangements to be put into place. However, these changes contemplated would remove a good deal of the risks currently blighting the industry and present a sound base for distributed ledger and smart contract adoption. All of the envisaged changes above can be coded and could be accommodated within smart construction contract arrangements.

This chapter has identified those initiatives and best practice instances, which have a direct bearing on the facilitation of later technologies including distributed ledgers and smart contracts. The justification tests in terms of novelty, intuition, scalability, alignment, incremental nature, accessibility, and purpose are all present in some way shape and form in the topics covered.

Chapter ten examines the most important stepping-stone in the pathway to technological advancement which is the phenomenon of BIM. However, as shall be discussed, it may also be possible to take an alternative route.

Notes

1. Allen, J. G. (2018) *Wrapped and Stacked: "Smart Contracts" and the Interaction of Natural and Formal Language* European Review of Contract Law 14(4) 307 at 313.
2. Farmer, M. (2019) *What is Automation Good for?*, Building Magazine 1 July.
3. Building Research Establishment Environmental Assessment Method launched in 1990 by Building Research Establishments sets standards for the environmental performance of building through the design and operation stages.
4. Duncan Reed presentation 7/7/2020 Constructing Excellence.
5. https://www.digicatapult.org.uk/news-and-insights/press/digital-catapult-weather-monitoring-bloodhound-land-speed-record.

6. Klein, R. (2019) *Payment in the Construction Industry-Where are We Now?*, SCL paper D220.
7. NBS National Construction Contracts and Law Report 2018, available at: https://www.thenbs.com/knowledge/national-construction-contracts-and-law-report-2018.
8. Francis Ho, Head of Construction, Olswang Solicitors, available at: https://acarchitects.co.uk/fac-1-what-it-is-and-what-people-are-saying/.
9. Farmer, M. (2016) *The Farmer Review of the UK Construction Labour Model*, CEO/Cast Consultancy.

10 Building Information Modelling (BIM)

BIM has been the most prominent advancement in the construction sector in recent years. It has lead to greater integration and collaboration amongst project members. However concerns remain over liability, the legalities of BIM and the reliability of the software.

BIM has been identified as the means to deliver the Government's Construction Strategy targets of lower costs, faster delivery, lower emissions and improvement in exports. This chapter will postulate that the pathways of BIM and smart contracts are not inextricably linked. It is conceivable that smart contracts can remain at a simpler, purely transactional level. However, the smart contract process will need to rely on BIM uptake, as the platform in the short-to-medium term and BIM take-up would appear necessary before wholescale contract automation stands a chance.

Currently, pockets of good practice exist. BIM has become a part of the common parlance in construction notwithstanding the limited evidence of its impact on the ground. Smart contracts appear as a logical extension to BIM whereby the contractual performance itself becomes automated. BIM is similarly linked to the partnering agenda given its ability to promote collaborative behaviour. The collaboration lobby has been keen to embrace data and digital leading to the statement in the Saxon Report that "what partnering needed to succeed was BIM." Taking this approach further, we will explore whether BIM is what smart contracts need to succeed.

10.1 BIM explained

BIM was a term first coined in 2003 in the United States. It is a misnomer given that BIM also applies to infrastructure and asset management. A better term is Digital Design and Construction. BIM replicates what other industries have been doing for a good while under the heading "product lifecycle management." At the heart of BIM is a three-dimensional (3D) model of a project going beyond conventional design information (Computer Assisted Drawing or CAD) by incorporating non-graphical data including cost and programming data.

BIM and its related initiatives are seeking to create an environment where working with technology is second nature in construction. The common currency for all technologies is data. Crucially, BIM increases the scope and speed of data exchanges, which highlight procurement and contractual questions as to who provides what data, when it is best provided, and how it is used and relied upon.

The original preferred definition of BIM was *"a digital representation of physical and functional characteristics of a facility creating a shared knowledge resource for information about it forming a reliable basis for decisions during its life cycle, from earliest conception to demolition."*[1]

BIM involves a widening suite of working methods that become possible or necessary when built environment industries move onto a digital basis or use artificial intelligence. BIM has much to offer the construction industry. The hub-and-spoke model described as an improvement on the linear arrangement could well be characterised as having BIM as the hub. It holds the key to:

- improved productivity,
- a better arrangement for time, cost and quality,
- easier access to project information,
- co-ordinated construction documents which remain up-to-date throughout the project and
- improvements to the environmental impact of a building.

BIM also brings into focus an improved access to information throughout a building's life, right through to demolition. It therefore chimes with the circular economy movement's goals. The model itself also has value as a replicable blueprint of how a building should be designed and managed.

BIM requires contractors, sub-contractors, lead designers, architects, project managers and designers to work together and share information. In many ways, these perceived benefits resemble those claimed for partnering and alliancing on construction projects. In practice, this may mean that BIM projects are particularly suited to a partnering- or alliancing-procurement model, especially for projects using BIM level 3 or beyond.

10.2 BIM levels

The state of readiness of any stakeholder within the industry is referred to as BIM "maturity." Surveys suggest that there is a good deal of exaggerated claims on how much the processes are being used and the level of sophistication achieved. Notwithstanding these statistics, a recent study concluded that there is a great lack of well-educated and trained BIM professionals. The most popular BIM software is Revett although there are a host of BIM-related tools in the market.

BIM represents a spectrum from projects using paper drawings to those using a fully integrated web-based database.

Level 0 – two-dimensional drawings, possibly created in Computed Aided
   Design (CAD) shared in hard copy format
Level 1 – using CAD together with a standardised approach to representing
   data across the construction team

Level 2 – data is presented in specific BIM databases which may also include information about cost or programming. The models are produced by individual members of the project team and remain "federated" and not yet integrated into a single model. Level 2 BIM was mandated for all public projects by 2016 by the UK Government.

Level 3 – this is where all data is held on an integrated, web-based system that can be accessed by all relevant members and is presented in a standardised format. This should include costing information, programming and lifecycle-facility-management information.

In terms of the data content of a BIM database and its representation, this is often categorised by reference to the types of information ("dimensions") it included. BIM-enabled projects may be described as:

- 2D, comprising simple, two-dimensional drawings
- 3D, utilising three-dimensional CAD to create a geometric representations of building components
- 4D, using 3D CAD and a database that includes information about the timing (programming) of the project
- 5D which incorporates cost information into 4D BIM
- 6D which adds information for facility management (FM) beyond completion of the construction project to 5D BIM

While it is not comprehensive, this way of categorising BIM maturity illustrates the extent to which BIM goes beyond 3D visual representation. The increasing dimensions of BIM provide an increasingly sophisticated platform from which to develop towards digital twin status as well as providing the infrastructure for smart contracts.

10.3 BIM through government initiatives

The Government Construction Strategy promotes the use of BIM. The Strategy acknowledged that at the industry's leading edge, there are companies capable of working in a fully collaborative 3D environment on a shared platform with reduced transaction costs and less opportunity for error.

The move towards BIM was given additional impetus by the construction strategy article published by the UK Cabinet Office requiring the submission of a fully collaborative 3D BIM (with all project and asset information, documentation and data being electronic) as a minimum by 2016 on public sector projects. This was initially regarded as a major step forward. Regrettably, four years on from the requirement, certain commentators find that there is limited evidence of real progress although inroads have been made. The biggest success has been around clash detection whereby the federated models are overlaid in order to spot inconsistencies between them. The irrefutable logic is that it preferable to resolve these errors at the planning stage than later on site.

The Government has also played an important role in creating a series of standards which sees the new ISO 19650 taking over from the PAS 1192 predecessor. The new international standard seeks to improve the exchange of information in formats, which prevent an overload. The emphasis is therefore on the delivery of information to meet the requirements and its careful management within a common data environment. Going forward, the ISO 19650 framework provides the foundations to enable machine-interpretable information to be exchanged by technology in a much more efficient way and thereby improving interoperability. It is hoped that the new international standard will add clarity to what can be an over-facing amount of detail required for BIM compliance in the short term. The risk is that the whole-scale change in nomenclature and expertise necessitated in the new approach could further de-rail the movement on BIM, as confused operators are sent back to the drawing board.

The Government policy of encouraging BIM use by demonstrating its effectiveness in public sector construction projects has provided a series of exemplar projects and developed alternative procurement models that are suited to BIM. Government Soft Landings (GSL) shares some of BIM's aims by encouraging early engagement at the design stage and emphasising the need for a smooth transition into the operation of a building. The Construction 2025 industrial strategy, launched in July 2013, refers to BIM as one way of achieving its long-term targets for improving the UK construction industry. The application of BIM level 2 has been identified as saving the UK Government £400 m a year, using the Benefits Measurement Methodology as trialled by Price Waterhouse Coopers on the Ministry of Justice Projects.[2] The Centre for Digital Built Britain's target of 33 percent lifecycle saving also shows how seriously the technology is being taken in the public sector.

10.4 BIM as the enabler

BIM is currently the expression of digital innovation within the construction sector. BIM is the main enabler for promoting collaboration and provides the framework on which other technologies, such as smart contracts, can improve the efficiency of supply chain transactions. As noted, the risk is that the opportunity BIM brings can be lost in the noise. Achieving these benefits depends on team members cutting through the hype, translating the jargon and aligning BIM with properly integrated procurement models. It is to be hoped that the new Framework Alliance Contract (FAC-1) achieves this and may succeed given the alignment of two-stage open book and FAC-1 in placing BIM at the centre of the arrangements. The enabling function of BIM is to help define, measure and manage the operational lifecycle cost by creating a direct data chain between design, construction, commissioning and operation of assets to enhance social outcomes and through data feedback mechanism provide a basis for continued improvement in asset design and performance. The interface with BIM and big data has also led to cost minimisation through energy optimisation and predictive maintenance.

BIM's establishment appears to be a pre-cursor to smart contracts in order to build a platform where the latter can operate. Smart contracts can be seen as an extension of BIM in that once the levels are complete then automation in the contract can occur. Fundamentally, BIM is about enhancing technological interactions through the creation of improved data as to design, cost, time and operation, all by reference to 3D models.

However, given a choice, clients will automatically favour a simpler option. Smart contracts offer this. Regrettably, BIM does not. BIM is complex for clients to understand and for their advisors to deliver. Three letter acronyms and initialisations proliferate and the number of people who understand, firstly, the contract and, secondly, the BIM procedures are limited to the experts. The BIM client is often unwilling to commit the time and cost needed to make the decisions required at the front end. The fanfare of BIM technology brought with it all the pomp and promise of a technology saviour that would transform the industry. BIM has struggled to make out its business case of generating savings for clients. The benefits are there, however, their intangibility and assurances that a longer-term view must be taken can act as a dampener on adoption. Put simply, BIM has become mired in its own detail. However, pockets of good practice and full engagement with the BIM mission exist and should grow. The number of professionals skilled in the art of BIM adoption will no doubt soon outnumber those that properly understand the complexity of established building contracts.

This situation can be resolved once the birthing pangs have passed. The hard work and dedication of those pursuing BIM deserve praise. Neither should one fall into the trap of casting around for the next big idea before properly examining the potential of the current. Nevertheless, the complexity of BIM is of concern given the limited attention span of funders, clients, and their short-term focus on cost savings and predilection for risk dumping.

Notwithstanding the above, BIM continues to receive a good press in the construction industry and unwavering support by the UK Government. The Farmer Report has BIM as a key deliverable of change within the construction industry.

The temptation to set off in a different direction to BIM ought to be resisted as far as possible. Smart contracts should be complimentary to the developments and build on them. The focus needs therefore to be on the incremental steps that facilitate adoption whilst addressing real and perceived barriers to implementation. This has been the approach of the FAC-1 that has sought to integrate BIM into its contractual network.

This is achieved by emphasising the importance of data transparency and team integration through direct relationships and direct mutual licences of intellectual property rights. There is further integration of BIM management with governance and clash resolution through Early Warning provisions as well as agreed software and clarity protocols for reliance on data. The BIM deadlines, gateways and interfaces become contractually explicit to ensure there status as key deliverables. The contract also has the flexibility to bring in BIM contributions from sub-contractors and manufacturers, occupiers, operators, repairers, alterations and demolishers.

The final piece of this integrated approach is the scope to capture the learning through the BIM provision with success measures to ensure continuous improvement. These improvements can be made by linking BIM to smart contracts in the following areas:

- Improving construction productivity: supply chains in construction result in high transaction costs that can be reduced through smarter contracts integrated with BIM. Not only are processes automated, but also by increasing levels of trust, the benefits of collaboration can be properly accessed.
- Rethinking procurement practices: traditional procurement practices have been criticised for many years, most recently in the wake of the Carillion collapse. The implications of smart contracts for commercial processes could represent a strong argument for procurement reform.
- Changing organisational culture: smart contracts and BIM reduce the potential for failure (deliberate or innocent) to operate construction contracts properly, thus reducing the potential for costly disputes.
- Increasing contractual transparency, assurance, and provenance for the benefit of contractors and suppliers in project supply chains. Many of the latter are SMEs and feel contractually disadvantaged yet add most of the value to the built asset. An automated contractual payment mechanism, in conjunction with a project bank account, would afford better protection from delayed payments, lost retentions and provide more payment security.
- An increase in confidence and trust may reduce contingency pricing, improving the conditions for the earlier and better participant involvement, and more effective collaboration to the overall benefit of the project and its owner.
- Interoperability of digital technologies: the importance of forging synergies between different but complementary technologies in BIM/FM and smart contracts is a key goal (Figure 10.1).

10.5 BIM issues

There is a stage in the adoption of a new procedure where the time comes to make something work. BIM has moved into this "hard miles" stage as the construction industry wrestles with the disruption it causes to its status quo.

The disruption is made the more acute by the lack of suitably skilled people to bring about the changes required. This is known as the digital skills gap. Around 90 percent of new jobs require digital skills to some degree and 93 percent of technology companies found that the digital skills' gap affects their commercial operations. The ability to check and verify that data is correct is currently extremely technical. Even with off-the-shelf BIM and FM products, individuals have to build their own logical rule sets. The resource is simply in very short supply in the construction industry.

The situation is complicated further by the perception about BIM that there is too much front effort required and not enough evidence of value delivered.

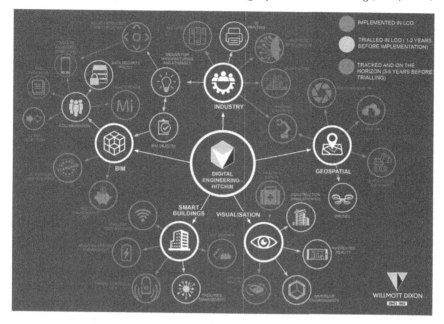

Figure 10.1 A pathway towards an integrated digital approach

Other problems experienced in the take-up of BIM include lack of compatibility between software platforms, confusion over ownership of the designs involved and the reliability and accuracy of the data.[3]

Therefore, it would not be accurate to portray the construction industry as fully in line with the BIM movement despite the claims of the National Building Specification (NBS) BIM surveys. These annual surveys (2020 was the 10th) collect the views of forward-thinking construction professionals who offer an optimistic appraisal of the penetration of BIM into construction. Awareness and adoption has grown from 10 percent in 2011 to 70 percent in 2019. The reports draw a picture of a two-speed industry with the BIM engaged and the BIM laggards. The limitations of the construction industry are the previous observations of not being good with change are evident here. The existing arrangements in the industry are all too often characterised by projects funded by third-party financiers concerned only with protecting their investment. The requirements for lending are expressed in terms of collateral warranties and step-in rights rather than any interest in 3D modelling and asset-management data.

Issues as to the reliability of BIM computer software programmes and models have encouraged a defensive contractual approach to legal liability. As has been seen, BIM enables design inconsistencies to be revealed through clash detection. The notion here is that issues arising between the different designers' inputs can be resolved in the model before it is built without the ensuing problems that would be encountered in real time. This gives rise to additional work in re-designing and the issue of whether this is treated as a claim needs to be addressed. Perceived threats to intellectual property rights are also a potential obstacle to the adoption of BIM. Ownerships and permissions should be clearly stated and licences should be granted in regard of contractor and sub-contractor involvement.

BIM level 3 renders the contributors work indistinguishable and does not notify the author of a change. This is where the multi-party future of construction contracts appears unavoidable. Only by having a joint policy of insurance can insurable liability be dealt with. PPC2000 has direct mutual licences. The BIM Execution Plan is where the roles and responsibilities of team members are set out. Making this document, together with the Employer's Information Requirements contract documents, appears to be the step forward in concretising BIM in the contractual process.

Surveys undertaken demonstrate that a single platform is not yet established as the BIM provider. Any problems experienced with the software and whom this exposes to liability is a concern. The interoperability of the software is another outstanding issue. Another issue surrounds the ownership of the Model and access to it. The NBS report that BIM is making headway and the legal ground-work is being put into place to facilitate level 3 take up. However, it is worrying that, in 2018, BIM level 3 is described as "yet-to-be-defined" (NBS 2018). This has to be of concern to the BIM supporters if there is to be any chance of a smooth and sure progression through the BIM levels.

10.6 BIM and contract drafting

On a BIM-enabled project, the building contracts and professional appointments will address BIM in:

- A BIM protocol, which is a distinct document setting out the practical ways in which BIM will be implemented
- Contract clauses which incorporate the BIM protocol as a contract document and, if necessary, deal with any other legal issues. The extent of any contract amendments depends on what BIM level is required.
- In addition, a professional appointment's schedule of services is likely to require amendment to reflect BIM-related services, although the extent of such changes depends on what services are required.

Standard form contracts have taken a light touch to inclusion of BIM provisions to date. Standard forms of construction contract refer to the Construction Industry Council BIM Protocol (CIC Protocol).[4] This umbrella organisation has sought to encourage BIM adoption by providing a contractual "add on" which operates by being executed by the client bilaterally with each project team member, including every consultant and main contractor. The network of bilateral agreements is completed by the main contractor signing the agreement with all its sub-contractors and suppliers making design contributions. The CIC

Protocol takes precedence over the consultant appointment, building contract and sub-contracts in the event of a conflict or discrepancy. The first edition was published in 2013, as the benchmark for BIM-enabled projects in the United Kingdom. The third version is now published. Reviewing the three versions provides a good illustration of the challenges and approach that has been taken to firstly encourage adoption through a light touch that has gradually been replaced with a harder edge.

The interconnecting network of bilateral contracts is an effective way of creating the necessary "consensus ad idem" if there is not going to be a multi-party contract. The multi-parry route is generally accepted as being the direction of travel. The PPC2000, FAC-1 and NEC4 Alliance Contract already use this approach. The contracts claim their contracts can be used to deliver BIM without any amendment.

The shortcomings of the current prevailing attitudes towards BIM are evident in the first edition of the CIC Protocol. The light touch is evident in that there is no warranty as to the integrity of the electronic data exchange and no liability for the modification, amendment, transmission, copying or use of BIM models other than for agreed purposes.

The first edition of the Protocol contained the obligation on the project team members to use "reasonable endeavours," which are recognised as a lower, less clear duty of care than the widely accepted standard of reasonable skill and care. The compliance with the timetable is also stated to be *subject to events outside the reasonable control*" described as a generic exception, which overrides the detailed provisions for extensions of time contained in most standard form building contracts. The wriggle room built into this approach was meant to allow people to get used to BIM in a way that will not heighten their anxiety about liabilities for either their board of governors or their insurers.

This was always going to be an interim position though and these limitations were partially removed in the second edition of the protocol, which adopted reasonable skill and care for the BIM obligations. The logic here is that BIM is now established as business as usual (BAU) and its obligations ought to be treated as seriously and as onerously as all other functions. This appears to be the correct approach. A third edition of the protocol will be necessary given the movement away from British Standard to International ones.

10.7 BIM best practice case study

UWE Bristol is an early adopter of BIM technology and a firm believer in the benefits it can bring. Its recent multi-million pound developments have set regional benchmarks in the BIM approach thanks to the contributions and enthusiasm of the project participants. The Faculty of Business and Law opened in 2017 and was constructed with BIM, and the learning was available to the students and region alike in the dissemination of results and the open access to the model itself. The Engineering Building (opened in 2020) will take BIM on to the further advances in terms of managing the built asset and returning some of the saving promised by BIM in the medium-to-long term. Plans are well developed for a complete digital campus featuring both existing and other new developments.

The Faculty of Environment and Technology offers a Masters course in BIM that has proved popular with students from around the globe. The course is not merely classroom based and features one module in particular, BIM in Business, where the students become BIM ambassadors and change managers by bringing BIM to local businesses. One such placement has recently generated a PhD opening, to explore the potential of this technology further.

The focus on this case study is not one of these recent projects using BIM. Instead, the example selected features term maintenance of student accommodation. This example has been chosen because of several advantages offered and existing synergies with a smart contract approach.

10.7.1 Inch-stone nature

Each one of the maintenance tasks logged and actioned is, in effect, a mini-contract. Smart contracts backed by the blockchain or other distributed ledger are comfortable with the volume of contracts generated. This represents one of the major benefits of a technologically enhanced approach. Each transaction can be logged on the system and the transparency and validity relied upon.

10.7.2 Non-critical nature

Maintenance contracts are not time/cost critical or cumulative in the same way as a standard building contract. Whilst targets are set for closing off job requests, there is no "bigger picture" and delay on a single item will not have the same consequences as delaying a critical path item on a building programme. In short, it is less risky to attempt to use smart contracts in this context.

10.7.3 Simplicity of coding the smart contract

A popular misconception in the area of smart contracts are that they are too difficult to code. This is not the case. The Accord Project has created a language called Ergo which most computer programmers can pick up easily. The issue is not "can it be coded?" but whether the coded contract, which is indistinguishable from a computer programme, can be embed satisfactorily in the mainframe of other systems

It is debatable whether smart contracts can cope with tests allowing for the exercise of discretion such as reasonable and satisfactory. The benefit of a maintenance contract task is that the contract will be simple and straightforward. The contract will come into being: "If X generates a work order then Y agrees to do it in return for the fee agreed." The performance of the contract will be acknowledged, say, for instance, by a university representative approved installation at the time of execution. The payment will be transferred – the university pays the installer directly.

10.7.4 Shortcomings in existing payment arrangements

It is at main contractor level that the nervousness of the future of building contracts is felt. When portrayed in an unflattering light, they can be depicted as the middlemen who unjustifiably sits on their supply chain's money. The latter are often depicted as the hard done to and victims of extended payment terms in a market place where insolvency often stalks the weaker, less-resilient members. There is some truth in this picture but the main contractors should take comfort from knowing that the clients need them. The client's core business is not the management of subcontractors and ensuring that the latter deliver on their promises. Main contractors are best placed to manage the supply community and will retain their usefulness here. The whole essence of the project bank account was to remove the non-transparent element of the main contractor's profit whilst ring fencing their ability to claim a decent price for a decent job. The short-term vision of the client focusing on the lowest tender price has been the root cause of the main contractors keeping this other income-generating source from their view.

In the new smart contract arrangement, the client can contract directly with the sub-contractors but will still need the intermediary of the main contractor to ensure performance. A system of cross-indemnities similar to the American mechanic's lien arrangement may operate here.

10.7.5 Shortcomings in existing inspection arrangements

Maintenance contracts routinely feature elements of technology-enabled solutions. Maintenance operatives regularly use their smart phones or iPads to take pictures of their work and upload them to the main contractor. Inspections follow in a small percentage of cases and the results of the inspections are extrapolated over the whole package of works. If a fault was found in the 5 percent of work examined then a similar percentage deduction on quality will be made across the board. This approach of "catch me if you can" is no different to time-honoured customs checks where the officers might cut open one bag of grain to examine whether the cargo is as per the bill of lading. Something more scientific is possible as we have entered the third decade of the 21st century. The client should be able to pay for exactly the service received at the point at which it is received.

10.7.6 How a smart contract approach could work with a BIM model

The student-accommodation block could have a BIM model that is a real-time record of its construction and operation. The "moving parts" of the building, say for example, the heating and ventilation system and automatic doors have their own sensors which feed information on their performance to the BIM dashboard. Such technology exists and is in current usage. A works request is generated by either a third party or the model itself noticing a low reading. The variable terms of the smart contract are automatically generated and automated execution follows. The contract is recorded on the ledger, as is performance when the operative logs completion of the task. The payment (also logged on the ledger) follows directly to the installer within minutes of completion being logged. The BIM model is updated with the working part now no longer listed as work to be completed. The main contractor would be on hand to source an alternative operative and to facilitate the performance of the task as required.

10.7.7 How a smart contract approach could work without a BIM model

The premise is used here that the case study involves the installation of heating ventilation and air conditioning (HVAC) equipment. This equipment is valuable enough to have sensors embedded within it. This allows the earned-value approach to be the main driver in the execution. Any reported malfunction or defect in the component equates to a reduction in the value of the unit. Less value if being derived, whether this is represented in student dis-satisfaction with the temperature in the bedrooms or against a performance target. The correction of the issue will re-instate the value of the asset and the smart contract execution and performance will be recorded as per the earlier example. The difference is that the approach is stigmergic rather than the hub-and-spoke approach.

The case for smart contracts to use BIM as part of their fulfilment is compelling. BIM provides the yardstick against which the smart contract can align. The execution and completion of each smart contract task can be referred to the model and actual progress can overwrite planned. The value extracted from each completed contract is therefore measurable and demonstrable. This inch-stone approach amounts to each mini-contract contributing towards the fulfilment of the project. The essence of the model is hub and spoke. Smart contracts could be viewed as the logical extension of the BIM levels. At this level, the BIM model is not simply the digital description of every aspect of the built asset but also of its execution and performance. The question is therefore – will smart contracts follow this route to BIM fulfilment?

The alternative is for smart contracts to avoid BIM and to progress on the purely transactional earned-value route. Here, the completion of each task is an end in itself, whether or not it is referred to the wider completion of the BIM model. The analogy is to termites completing their pre-destined task. Each termite knows not what the others do and yet the mound is built. The mound is built on the inch-stone, as opposed to milestone, principle using a distributed or stigmergic approach (Mcnamara 2017).

It is open to conjecture as to quite how soon the construction industry will reach sufficient maturity to embed a data-led inch-stone approach. The availability of powerful handheld computers in everyone's pocket in the form of smart phones is an obvious starting point. These terminals permit the upload in real time to a programme that continuously overwrites planned with actual to demonstrate value. The embedding of censors in devices is already in wide usage and is set to pass 25 billion by 2020. Whether this means censors in every brick space or capping stone remains to be seen. Heating and ventilation units have already multiple sensors recording performance and maintenance issues. The earned-value model offers a different sort of control and overview, which will be automated to a degree previously unseen. The project manager can observe the progress without needing to run the diagnostics checks on performance and completion in the same way.

A popular view is that the construction industry requires a disruptive influence in order to force it down the route of digitisation and engagement with the cutting edge of technologies in other sectors. This is not necessarily the case given the platform offered by BIM. Smart contracts are a complimentary technology and ought to explore both paths presented by the fork in the road. Smart contracts might be what BIM needs to succeed. Equally, smart contracts can be at the forefront of the disruptive intervention apparently required.

10.8 BIM into the future

For the pioneers, BIM is seen as business as usual and is taken way beyond its current perception as being for designer's only. The potential to incorporate GPS machinery and trace and predict literally every nut and bolt of the process.

One of the most complimentary technology for BIM is augmented reality. Previously, BIM appeared destined to interact with virtual reality – the client could enjoy a virtual tour of the new building and make design decisions about where certain items should go – where would the coffee machine work best? However, the greater use in construction to embed BIM through the whole process is in overlaying actual spaces with the models of components and buildings that have been assembled. Audits become so much more routine where the certifier can view the installation through an augmented reality head set, which overlays the precisely detailed plans on what they are actually seeing. This is taking the BIM to another level of site sophistication and, if adopted, would result in BIM as BAU.

Greater strides towards improvements in BIM come from the common platform and the ability to have the content enabled across all formats with embedded data you can see or conceal as you wish. This can involve an exchange between the model and the actual data being produced to provide a fully up-to-date virtual record of the building asset. The capability is then extended further into machine learning opportunities for the model to analyse its own performance and optimization through algorithmic adjustment. This is sometimes referred to as generative design. This is the desired outcome that gives architecture, engineering and construction professionals insights to efficiently plan, design, construct, and manage buildings and infrastructure. In order to plan and design the construction of a building, the 3D models need to take into consideration the architecture, engineering, mechanical, electrical and plumbing (MEP) plans, and the sequence of activities of the respective teams. The challenge is to ensure that the different models from the sub-teams do not clash with each other. The industry is trying to use machine learning in the form of generative design to identify and mitigate clashes between the different models generated by the different teams in the planning and design phase to prevent rework. The software uses machine learning algorithms to explore all the variations of a solution and generates design alternatives. It leverages machine learning to specifically create 3D models of mechanical, electrical and plumbing systems while simultaneously making sure that the entire routes for MEP systems do not clash with the building architecture while it learns from each iteration to come up with an optimal solution.

Building managers can use these functions after the construction of a building is complete. BIM stores information about the structure of the building. AI can be used to monitor developing problems and even offers solutions to prevent problems.

10.9 BIM and digital twins

An extrapolation of what is achievable in relation to a single-built asset leads to the notion of a complete digital version of the built environment. Digital twins can be seen as a logical extension of BIM at an individual and national level. A digital twin is a digital replica of a physical asset, which creates digital simulation models that update and change as their physical counterparts change. A digital twin continuously learns and updates itself from multiple sources to represent its near real-time status and working position. Urban planning practice has seen an increasing appetite for digital technology in the Smart Cities movement to model urban environments.

The work of Digital Build Britain is to bring about the combination of BIM will the internet of things, advanced data analytics and the digital economy to allow more effective planning, lower-cost building and operation costs. Digital Build Britain is the successor to the UK BIM Task Group and is a government-funded body.

The Digital Twin movement relies on the Gemini Principles targeting the public good and alignment for stakeholders throughout the built environment. These laudable principles also cover openness, federation and security. Other key publications include the Institute of Engineering and Technology from 2019 publication and Data for Public Good from 2017.

A fully responsive, automated holistic system for entire countries is recognised as being something of a unicorn. However, small steps as this book as expounded, add value and will only increase as technology and techniques improve. The Digital Twin can operate as much more than a BIM platform as a data resource that can improve the design of new assets and the understanding of existing asset condition. The twin can verify progress during a build and run "what if" simulations and scenarios.

The term digital twin first appeared in the early 2000s but the first recorded use was on the Apollo 13 rescue mission where NASA referred to it as a "mirrored system." Science fiction and space travel facts combine poetically in this passage of history now 50 years old. The concept has since developed into an integral digital business decision assistant and a bridge between the physical and digital.

Table 10.1 Digital twin maturity spectrum

Voluntary	Mandatory
Negotiation Mediation Dispute Review Boards	Adjudication Arbitration Litigation

A maturity spectrum (Table 10.1) is proposed by the Institute of Engineering and Technology (IET) to communicate progress. The spectrum recognises that the purpose and value of increased complexity and connectedness are clearly identified, justified and realised.

* 0 – Reality Capture (e.g. point cloud, drones, photogrammetry)
* 1 – 2D map or 3D model, object based only (clash detection)
* 2 – Connect to static data and BIM
* 3 – Enrich with real-time data (give right time decisions)
* 4 – Two-way integration and interaction – control the physical from the digital
* 5 – Autonomous asset with operation and maintenance, self-governance

Element 0 Reality capture is already in use for as-built surveys and fly arounds for hard to access or dangerous sites. Element 1 mapping and modelling allows for design optimisation and co-ordination. This represents the high tide of many existing BIM arrangements. Element 2 enriches the federated BIM platform with data, such as reliable information on the time and cost simulations for the build. Element 3 interjects real-time data and leads to operational efficiency through smarter facilities management. The really clever stages start at Element 4 with two-way data integration and interaction. For example, an operator could manipulate a physical valve from the twin. This level of integration requires a high level of additional sensor and mechanical augmentation of the physical asset. The final element is cited as autonomous operation and maintenance with total self-governance. The goal would have been achieved here of machine learning and little or no human interaction required. In a rather chilling statement, the IET recognise:

> The full ramifications of what Element 5 maturity means, and the quantifiable benefits it will bring, are yet to be fully understood.

And yet, the potential to do good through the medium of digital twins make a compelling case. The movement is enabled by connectivity, 5G storage and capacity. Retrofitting of the current built environment stock can reduce carbon emissions by 80 percent by 2050. Compiling a digital twin for existing assets allows for performance testing and for confidence and reassurance to be provided in the net effect of the retrofits carried out.

The level of integration between digital twins is another area of huge potential. It is not too difficult to envisage a national infrastructure with integrated industries such as waste/sewerage, water telecommunications, transport, environment agencies and energy networks. These systems could be in constant dialogue to optimise the built and natural environments for the town/city/country/world population. It is only by maximising the use of latest digital technologies and construction methods that we have any chance of meeting the incredibly onerous challenge presented by the forecasted population growth. Perhaps, we should have a digital twin of Mars, just in case.[5]

10.10 Conclusion

BIM has been a force for positive change in the built environment sector and has increased awareness of and receptiveness to wider-scale digitalisation. BIM when operated properly it can lead to a reduction in waste, mistakes and disputes on building projects. BIM continues to receive unwavering support by the UK Government. Construction 2025 names BIM as a key deliverable of change within the construction industry.[6]

BIM serves as an essential data source that smart contracts can rely upon for their operation, and can accommodate different view definitions and can communicate physical and semantic information that drives built environment applications. For example, facilities management digital management requires information in the form of data triggers from the appropriate BIM model. Those same data triggers are essential for the development of smart contracts.

BIM on site can provide a powerful resource to ensure that the gap is bridged between the mental model of the design team and the reality that everyone can appreciate through the supply chain and down to the end users. The value in the information is also in the provenance-based approach to procurement which is likely to grow in importance in the future. As climate change becomes one of the defining issue of our time, so too does the need to establish where the components have come from and their sustainability credentials. This will impact not only on the original occupier of a building, but also the subsequent ones. The provenance and performance of the building will increase in importance and may, in time, be as important as the title documents in terms of that building's marketability and social worth. Information, in the form of data, can allow the integration of science based decision-making with technology-focused product procurement choices.

The instigators of BIM cannot have foreseen an alternative route to automate and digitise the industry as is now available through distributed ledger technology. The smart contract movement is a newer, fresher concept, which can decide whether to support existing technologies or write its own narrative. It will doubtless soon be superseded as the celebrated cause of the day. The benefit of compatibility, both with antecedent and later technologies is a hallmark of permanence and resilience. This could serve smart contracts well in the construction sector.

Future gazing is an imprecise and difficult art. This author has sought to draw together some strands in an effort to predict future development. Whether or not such fashionable debates prove entirely academic remains to be seen. The challenge remains for policy makers and lawyers to recognise the demand for a new contractual response and to deliver this in the most expeditious and efficient manner possible.

Notes

1. Snook, K. (2018) Drawing is Dead – Long Live Modelling, Construction Project Information, available at: http://www.cpic.org.uk/publications/drawing-is-dead/ last accessed 10/07/2020.
2. http://www.bimplus.co.uk/news/bim-benefits-report-government-could-save-400m-yea/.
3. Mason, J. (2017) *Intelligent Contracts and the Construction Industry.* Journal of Legal Affairs and Dispute Resolution in Engineering and Construction 9 (3) .
4. https://www.cic.org.uk/download.php?f=the-bim-protocol.pdf last accessed 10/07/2020.
5. Billions of years ago, the resemblance between Mars and Earth was quite close– both were wet and warm although Mars is only half the diameter of Earth.
6. Department for Business, Innovation & Skills, *Construction 2025: Strategy*, GOV. UK (2013), available at: https://www.gov.uk/government/publications/construction-2025-strategy.

Section V
Online dispute resolution and smart contracts

11 Online dispute resolution

11.1 Introduction

There remains one area for this book to cover which completes the review of digital developments in construction law. The aim is to return to the starting point of the discussion on the construction industry and its limitations, in particular, the high incidence of disputes. Disputes are the first thing that spring to mind when the subject of construction law is addressed. Dispute in construction is essentially viewed as an inevitability under current practice and, as such, the contract is considered to be the rules of engagement. Lord Donaldson summarised the point thus:

> I cannot imagine a (civil engineering) contract, particularly one of any size, which does not give rise to some disputes. This is not to the discredit of any party to the contract. It is simply the nature of the beast. What is to their discredit is if they fail to resolve those disputes as quickly, economically, and sensibly as possible.[1]

It would be rewarding to think that this was not the case in the future and that the developments outlined in this work had gone some way to counteracting the current situation or at least presented some opportunities for another way. However, we humans generate billions of disputes each year, soon to be tens of billions, as the growth shows no signs of stopping.

There are a number of threads to pull together from the work and the subject of disputes along with current initiatives in the law to review and align with the wider context. This chapter reviews the opportunities and challenges that an online dispute-resolution system might bring, what it entails and its potential impact.

The enduring nature of disputes in the construction industry is no doubt a cause of some bafflement to would-be smart contract writers. To their mind, disputes should not exist as the computer code would be infallible and incapable of giving rise to a dispute, after the initial teething issues. The granularity provided in an inch-stone approach to construction would mean any false step would be easily reversible and that no legal infrastructure is necessary. This prediction does not give adequate regard to the contentious nature of business where mistakes and money are concerned. A built-in species of dispute avoidance and resolution will ultimately be necessary. The consequences will not all be expressible in code. There will be room for legal dispute. There are likely to be claims in mistake and misrepresentation and arising from coding error.

However, the new terms of reference for construction implicit in the digital approach, which has been outlined, does offer some new frontiers for dispute resolutions. It is possible to expedite dispute resolution by entrenching processes and procedures within the code itself before ultimately allowing for a form of oracle decision-making or alternative dispute resolution (ADR) or judicial resolution.

The greater granularity referred to should have a positive impact on removing the triggers for dispute. Of the three key outcomes of the digital movement – collaboration, transparency and traceability – it is the second and third that should vastly reduce the incidence of disputes. Auditability of the events as they happened and an immutable record of events provides a (hopefully) incontestable single source of truth for what occurred. Further, the advances in payment provision down the project bank account type arrangement would reduce the volatility and risk to organisations over asset ownership and insolvency. That said, it would be naïve to deny that many disputes are due to other factors such as poor business planning, personalities and a simple inability to admit to oneself or one's financial backers that the situation is not as per the predictions in a set of accounts.

Nick Szabo's original work on smart contracts included some contemplation on in-built remedies for contract violations. The opportunity was highlighted to impose immediate sanctions for breaching contract terms such as a default on a loan automatically imposing a higher interest rate further following a series of warnings and/or revoking licences to use certain software interfaces. Another example was a hire car locking its driver out if the terms of the hire contract were not met. This could be updated further in the wake of autonomous vehicles by the driver suffering the ignominy of watching the car drive itself back to the depot. Traditional contracts rely much more on a judicial system that decides on punishment for breach and enforcement after the event. Automating this reactive process should reduce the incidence of claims although it has ramifications for consumer and basic rights in terms of the right to a fair trial and defend one's position. In these cases, there has to be protection by way of appeal against smart contracts executing in an unexpected manner in order to build trust in the system. The auto-correct function might be available through the entity responsible for creating the smart contract in the first place. This could be a "software as a service" approach to effective policing of the smart contract operation. In other cases, where the contract behaves in an unexpected manner, but the party who created it is unable or unwilling to correct this behaviour, some sort of arbitration mechanism will be required to settle the dispute. This is where different types of oracle might be involved depending on the nature of the issue – is a correction of data entry required or potential a ruling by a semantic oracle on the different interpretations possible of the natural or computer language of the smart contract.

This chapter follows the lay out of the rest of this work in terms of describing the position AS IS, TO BE and the STEPS INBETWEEN. The AS IS position involved the current crop of dispute-resolution choices and their emphasis on avoiding the excesses of litigation where possible. The TO BE discussion looks briefly at some of the novel approaches of blockchain and smart contracts, including the possibility of using the technology itself to be the dispute resolver. The STEPS INBETWEEN focus on online dispute resolution (ODR) as a stepping-stone between the two other positions and the most likely route as the enabler for future progress.

11.2 Dispute resolution, AS IS

11.2.1 Range of existing dispute resolution provisions

Disputes arise for many reasons including technological and ideological differences combined in a mixture of complex requirements, multiple disciplines and competing interest; inaccurate and inadequate design information, inadequate site information, late decisions by client, poor communication, unclear risk allocation, inadequate communication, and/or deficient or ambiguous contract documents (Figure 10.1).

Two international consultants, Arcadis and HKA, have conducted separate studies on the subject of disputes within the construction and civil engineering sector and have found common areas of agreement; however, the number one cause for disputes varies. From the research conducted by Arcadis, over the six-year period (2012–2018),[2] there have been three consistent reasons for dispute which can be ranked as follows:

1 Failure to properly administer the contract
2 Poorly drafted or incomplete and unsubstantiated claims
3 Owner/contractor/contractor subcontractor failing to understand and/or comply with his contractual obligations

There exists some commonality with regards to the causes of the disputes. The causes identified by Arcadis have been repeatedly documented over the years and appear to have easy fixes, yet they continue unabated and are the source of millions of pounds in disputes and legal expenses on an annual basis. The central failing, a lack of planning and unwillingness to stick to the agreement, does not bode well for smart contract adoption that is based on a meticulous approach to pre-commencement activity. The assumption must therefore be made that the necessary investment in time and money will be made for the rewards on offer, most importantly an end to the very disputes on which some businesses depend.

Figure 11.1 Dispute avoidance and resolution

There are many processes that can be utilised for dispute resolution in the construction sector. The most popular are negotiation, mediation, adjudication, arbitration and litigation. These forms of resolution are being utilised internationally throughout the construction industry. The following are some of the characteristics of the various forms

- Negotiation – a party-to-party method that is not binding
- Mediation – the use of a mediator (a go between) who tries to settle the differences, decisions made are non-binding unless any agreement is signed by both parties
- Adjudication – this is a statutory right to a short form of binding decision-making which can be used for any construction dispute
- Dispute Review Boards – this is a pre-selected board of experts are involved in the project during the build and manage any claims and disputes arising
- Arbitration – a process which requires a third expert party to decide a case following full submission and, due process, the procedure is expensive and lengthy
- Litigation – this comprises the court system representing access to justice for individuals and businesses Litigation underpins all the other dispute resolution choices as the final enforcement choice.

11.2.2 Negotiation

Negotiation between parties is the most common or preferred method of resolving disputes. The level of formality and medium by which this is conducted vary widely depending on the context. This can range from a high-powered board room meeting between executives to a "clear the air" sit-down to discuss a payment application on site. The important features is that this is non-binding and simply requires the parties to attempt to come to a mutually beneficial accommodation. Negotiation requires the foresight to recognise that escalating the dispute is not in anyone's interest apart from the lawyers. Personality clashes can frequently prevent progress and can see other forms of dispute resolution being brought in to unblock an impasse.

11.2.3 Mediation

Mediation is a voluntary, non-binding, private form of dispute-resolution process whereby parties are assisted in their negotiation by an independent third party, the mediator. The mediation typically involved some shuttle diplomacy by the mediator who uses various tools and techniques to bring the parties closer together on the issues in dispute between them. The parties can be brought together at the start and the end of the process when hopefully the final stages of negotiation will be concluded and a binding agreement signed to capture the terms of settlement arrived at.

The process involves the preparation and disclosure of a position statement which is a without prejudice opening to the procedures. Without prejudice is a term meaning that the proceedings and any records of what was disclosed cannot be used in subsequent legal proceedings unless both parties agree. This means that a party's formal pleaded case in another form of dispute resolution, say litigation, is not "prejudiced" insofar as it will not have been compromised by these voluntary attempts to resolve the dispute. This enables negotiations and mediations to happen in an environment where the parties can be frank with one another and explore settlement without fear of consequence.

Mediation is not a judicial process and is performed on a without prejudice basis. There are some procedural norms such as position statements and opening addresses. It is a party-centric approach, requiring a short time period, typically half a day or two, and cost efficient. The parties control the decision-making and solutions can be creative and flexible. Parties have the opportunity to state their grievances and the records are confidential.

While mediation is a voluntary process, refusal or failure to take part in a requested mediation may be accompanied by cost sanctions in any subsequent litigation. These cost sanctions have become progressively more stringent for anyone not attempting Alternative Dispute Resolution (ADR). A pro-ADR approach has become an integral part of our litigation culture in the United Kingdom, with pre-action protocols, the Civil Procedure Rules and the Jackson reforms all promoting the use of ADR whenever appropriate.

The case of Thakkar v Patel[3] demonstrates the risk of ignoring a request to mediate. The claim involved a dilapidations schedule requiring payment of £210,000 that was met with a counterclaim for around £40,000. Both parties initially expressed a willingness to mediate by requesting a stay to ADR in their allocation questionnaires. The Claimant went on to attempt to make arrangements for a mediation but failed to hear back on this by the Defendants on several occasions. The ADR stay was lifted and the trial took place. The Claimants were awarded £45,000 and the Defendants £17,000 leaving a balance owing to the Claimants of £28,000. This left the question of costs outstanding, by this stage far greater than any of the sums in dispute. The Defendant was ordered to pay 75 percent of the claimants' costs whilst recovering their costs of the counterclaim. The Defendant's silence in the face of the offer to mediate ensured they were effectively penalised. The Judge observed that a pure money claim of this nature would definitely have been resolved with the help of a skilled mediator. However, had the parties not reached an agreement then there would not have been any cost consequences as the point is to try it.

Although it is considered non-binding, once the parties have arrived at an agreement, the agreement can be written up and signed by the parties. The potential for smart contracts would be to incorporate any such agreement either within their existing smart contract or to replace the latter with an agreement compliant version.

11.2.4 Adjudication

The right to adjudicate stems from the Housing Grants Construction and Regeneration Act 1996 following the recommendation for the same contained in the Latham Report. The observation that the supply chain in construction was being starved of its cash flow which often resulted in insolvencies as parties waited and waited for their hard-earned cash. Sir Michael recognised the need for a quick fire form of dispute resolution to allow parties to receive their entitlement. The visibility of what they were owed and why deductions were being made further empowered the supply chain. The adjudication process of dispute resolution remains relatively quick as the parties employ an impartial third party to reach a decision. The adjudicator is usually appointed by a nominating body in the absence of a named party being identified in the contract.

The process of adjudication involves the service of a Notice of Adjudication with the disputed contractual party and with the Adjudicator Nominating Body where relevant. All relevant documents pertinent to the Referring Party's claim must then be forwarded to the appointed Adjudicator, who sets a timetable for the remaining stages in the Adjudication. The Responding Party has a short time in which to prepare and lodge their Response with supporting documents. The Referring Party may have an opportunity to Reply to the Response before the Adjudicator then makes a decision based on the cases disclosed.

This quick-fire approach to dispute resolution is a blue-print for ODR and crowd-based approaches to resolution. The emphasis is on closing down the pleading stage and documentary evidence in support as quickly as possible to allow the dispute resolver the opportunity to see "all four corners" of the dispute in a single place. The ongoing nature of arbitration and litigation through the different stages of witness statements and expert reports have long identified been as adding to the time and cost of these more formal arrangements.

In the United Kingdom, statutory adjudication became available on 1 May 1998, with the enactment of the Housing Grant, Construction and Regeneration Act 1996 (The Construction Act) (HGCRA), s108 (1) and amended on 1 October 2011 to incorporate the provision that it was no longer necessary for a construction contract to be in writing. The purpose of the Construction Act emerged as a remedy to cash flow problems. The "pay now, argue later process" ensures that the project continues without interruption thereby earning the term "rough justice;"[4] the decision is enforceable and binds the party unless overturned by agreement, arbitration or litigation.

The courts have been supportive of adjudicator's decisions in the enforcement proceedings that usually follow on from a non-payment by the losing party. The summary judgment procedure and other interlocutory processes are used to ensure that the winning party is able to enforce payment. There are many instances where the Judges of the Technology and Construction Court have found procedural errors in the "rough justice" approach and refused judgment. However, on the substantive grounds, the courts remaining willing, to uphold the adjudicator's decision regardless of errors in fact or law. Examples of procedural flaws include lack of jurisdiction *Carillion Construction Ltd v Davenport Royal Dockyard*[5] bias or breach of the rules of natural justice.

Several standard contracts such as the Institute of Civil Engineers (ICE), International Federation of Consulting Engineers (FIDIC) and International Chamber of Commerce (ICC) have all published procedural rules for adjudication. The timeframe for the decision may vary; for instance, the Construction Act allows for 28 days with possible extensions. Adjudication schemes based on specific legislation are also found in Australia, New Zealand, India and Malaysia. Kenya, South Africa and Mauritius have or are in the process of enacting similar legislation.

11.2.5 Dispute review boards

A dispute board or dispute review board (sometimes also a dispute-adjudication board) is a site-based dispute-adjudication process. This typically comprises three independent impartial persons selected by the contracting parties. The significant difference is that the Board is appointed at the commencement of the project before any disputes arise and, by undertaking regular site visits, is actively involved throughout the project. This has been found to influence the performance of the contracting parties in a positive manner and has "real time" value. This early appointment and real-time usefulness of the Board has been borrowed as ideas in the design of ODR and blockchain-based resolution.

The right to appoint a Dispute Board stems from the contract itself, which should also take a position on the binding nature of the recommendations issued by the Board. A typical position is to allow either party to raise a Notice of Dissatisfaction with the Board's decision, which will lead to an escalation to the next dispute-resolution step identified in the contract. However, the recommendation is itself made available for the latter tribunal to see. This means that the dissatisfied party will have the additional challenging task of effectively demonstrating why the Board came up with the wrong answer notwithstanding their expertise and involvement in all stages of the project.

FIDIC[6] recommended a dispute-adjudication board (DAB) in the event of disputes and allows for a 84-day period, from referring the matter to the DAB for a decision. In its most recent 2017 revision to its contracts, FIDIC has replaced the DAB with DAAB, which refers to the Dispute Adjudication/Avoidance Board.

11.2.6 Arbitration

Arbitration is a more complex means of dispute resolution than the other methods previously mentioned; it is costly, lengthy and similar to litigation.

The main advantages of arbitration over litigation are its privacy and the proceedings are usually confidential; in highly technical matters, suitable arbitrators can be selected and appointed by the parties with cross-border disputes, the arbitral awards can be enforced in other countries who are party to the New York Convention 1958.[7] This reciprocal treaty is widely used around the world to the extent that being a non-signatory has a detrimental effect in any country's ability to attract foreign investment. The enforceability of arbitral decisions in this manner is an essential lubricant to global trade infrastructure and is the single reason why arbitration is the only serious choice for large and complex construction disputes.

Arbitration differs from litigation in the following respects. The parties can determine the rules to govern the procedure as opposed to court proceedings. The parties dictate the size of the arbitration panel whether a single arbitrator or multiple. In the case of a panel of multiple (usually three) arbitrators, each party selects one member of the arbitral panel of three. Those two selected arbitrators nominate the third and final one, who acts as the chair and cast the deciding vote if the two are unable to agree. The decision of an arbitrator(s) is final and binding with limited grounds for appeal; however, arbitration does not completely preclude court proceedings.

Arbitration may be an *ad hoc* process or governed. In the *ad hoc* process, the parties may determine the particular rules that they consider appropriate for the arbitration.In the governed process, the parties can select from a number of arbitral organisations and conduct the proceedings under their set procedures.

The arbitration process required several stages that can be broken down into:

- The agreement to arbitrate;
- The appointment of arbitrator;
- Planning for the hearing;
- Parties present their case and
- The award and enforcement.

Arbitration in England and Wales is governed by the Arbitration Act 1996, which defines, at Section 1, the object of arbitration is to obtain the fairest solution of disputes by an impartial tribunal without unnecessary delay or expense. The parties were stated as being free to agree how their disputes are resolved and the courts discouraged from intervening in the process as far as possible.

An arbitration agreement may be a clause within a contract agreement or a free-standing agreement in which the parties agree to resolve disputes by arbitration rather than by litigation. The decision of the arbitrator/panel is a written award which is binding on the parties and enforceable. The commencement of arbitration proceedings varies under the organisations and the major international institutions.

In considering arbitration as an option, the choice of the seat of arbitration is important as the laws governing the substantive issues will be the country of the project and the procedural rules of the arbitration will be governed by the laws of the seat. In addition, the court of the seat will hear any challenge to the award.

The main complaint concerning international arbitration has been its increasing cost, with the largest cost element typically being legal fees. An average arbitration may require between 1,500 and 4,500 hours of legal work. The Chartered

Institute of Arbitrators conducted a survey of 254 arbitrations conducted between 1991 and 2010, and found the average cost to be £1,580,000 for Claimants and approximately 10 percent less for Respondents.[8] Arbitrator fees and administrative expenses represented 20 percent of the total costs.

11.2.7 Litigation

Litigation underpins the other dispute-resolution procedures and is the court of final recourse. Access to public sector judges hired by the state are a cornerstone of society throughout the world. In some countries where no provision is made for any other form of dispute resolution, it remains the only choice. Whether or not, this is a viable choice that depends on the context. Brazil's judicial system faces a massive backlog of cases. In 2014, one group of five judges in Sao Paulo was reported to be handling 1.6 million cases.[9] In 2016 in India, there were 21.3 million cases currently pending in various courts in India.[10] One newspaper article express the problem thus: *"if the nation's judges attacked their backlog non-stop with no breaks for eating or sleeping and closed 100 cases every hour, it would take more than 35 years to catch up."* It is straightforward to imagine the benefits of online dispute resolution cutting through this backlog which blights access to justice in many parts of the world.

11.2.8 Conclusion

There is a massive opportunity for newer forms of dispute resolution to improve on the shortcomings of the construction industry and society in general, and its dispute-resolution procedures, in particular. The staggering cost of arbitration/litigation and the lack of access to litigation in some countries mean that ODR and other approaches such as crowd-sourcing justice can potentially replace all of these traditional approaches to dispute resolution in terms of resolving them more expeditiously as well as seeing the technology reduce the total number of disputes. . .

11.3 Dispute resolution, TO BE

It appears logical to conclude that the more computers continue to improve their processing power and ingrain themselves ever more deeply into our daily lives then the more likely that they will be better placed at resolving our disputes than we are. Neither would the approach taken by artificial intelligence (AI) would be constrained by a human approach. The AI merely need access to the data in order to formulate its approach. Data, as has been described, can reside in the blockchain or distributed ledger.

Blockchain continues to ride high on the hype-cycle and generates enthusiastic recommendations for its infallibility, transparency and immutability. Disputes are routinely overlooked in this exciting stage of development meaning that redress may well prove difficult to come by if it is not addressed. Disputes are so very ordinary, particularly in construction and there appears to be an inevitability that they will continue on the blockchain.[11] Disputes involving a fraud, mistake and loss of passwords are foreseeable along with the less predictable consequences of doing business in a fast-paced changing environment awash with inflated expectations. The message to be conveyed to the blockchain devotees is that there is an inevitability in conflict and that a trusted dispute-resolution procedure is required. This must allow for and respect the right of the individual not to follow a herd mentality on the capabilities of the technology.

Users of the blockchain have, in time, started asking the question about what happens in the event of a dispute. Many entrepreneurs have sensed an opportunity to develop different forms of ODR allowing for a flexible design and a global reach. One of the most interesting aspects of these platforms is the use of "the wisdom of crowds" effectively crowd-sourcing the outcome of the dispute via jury voting. An obvious limitation in this is only binary resolution is possible in favour of one party and against the other. The following companies are examples of providers of this and other approaches:

- Kleros "arbitration": This process can be activated once a dispute arises in the execution of a smart contract and freezes fund transfers until resolved. The Kleros system must be pre-selected and written/coded into the smart contract together with basic features such as which process applies and the number of jurors required. Jury members are incentivised by blockchain tokens. The aggrieved party approaches Kleros and send all the evidence it has through an encryption service. Jury members are anonymous and nominate themselves. After weighing the evidence, each juror commits to voting for one of the parties and must justify their decision. All votes are counted and the smart contract is executed in accordance with the decision that represents the highest number of votes. Tokens are redistributed among the jurors depending on whether they voted with the majority or not. The presumption here is that if the juror voted with the minority that the juror in question had insufficient expertise in the subject. There is a right to appeal which results in the number of jurors being doubled provided the appellant deposits the funds to cover the appeal.
- Juris provides a similar code for adoption in smart contracts. Once a dispute arises, the transaction is frozen and the Juris dashboard is accessed, which contains a range of techniques including mediation that is seeking to assist the parties to a consensual agreement. If this fails then resolution can be sought from neutral jurors operating a token system similar to Kleros. A distinguishing feature here is in relation to complex disputes where more experienced jurors can provide an enforceable judgment binding under the New York Convention. This is a more costly option but the importance of a globally enforceable decision is difficult to overstate.
- Sagewise can also be incorporated into a coded contractual clause as an equivalent to the traditional dispute-resolution clause. The novelty in the Sagewise approach is its ability to allow for the enforcement of any resolution

reached through its contract amendment feature. If the parties agree, the smart contract can be amended through the mobile app and then performance can continue on the new terms. If the parties do not agree then the next stage involves a human third-party facilitator who advises on the choice of dispute-resolution fora available. Sagewise is effectively a gateway to an ODR marketplace. The provider, once appointed, is granted full control over the smart contract and can enforce the decision reached by creating a new, decision-compliant, smart contract.

There is impressive ingenuity displayed in using the in-built features of the blockchain and smart contracts to resolve disputes through democratic decision-making and automatic enforcement or resolutions. Limitations exist in some of the approaches as they diverge considerably from the established cultural and legal barriers. That said, the all-important element of trust – in the system, its processes and fairness – could be generated from the approaches described above. Something is lost, however, in the anonymous nature of the resolutions on offer. On higher value and complex disputes, the resolver would probably place considerable importance on being able to "see" or form a view on the integrity of the parties, particularly where issues of conduct are concerned. Replacing this cultural norm with an anonymous approach will take some time to sit easily with existing quasi-judicial preferences.

The fast-evolving distributed environment operating across national borders presents a challenge to regulate. Some national regulators remain seemingly ignorant of the new technologies whilst others embrace it. The anonymous and pseudonymous exchanges represent an obstacle to traditional forms of dispute resolution and enforcement. However, a sophisticated approach employing measures such as those developed by Kleros, Juris and Sagewise amongst others, can lead to real-time interventions into the operation of smart contracts and their amendment to respect a decision reached. The prize of eliminating the need to chase enforcement is an enticing one.

A further tantalising possibility is the use of AI in dispute resolution. The "fourth party" is a metaphor that refers to the power that technology can add to the dispute-resolution process. The idea being that alongside the two human disputants and one human dispute resolver, there is a fourth party who plays an increasingly important role in the dispute initially through ODR but ultimately taking the role of dispute resolver itself. This concept of utilising technology in dispute resolution is the futuristic perception that a dispute-resolution process will utilise AI for decision-making or an outcome is not that farfetched as it once seemed. The popular imagination is seized by the idea of a digital judge. The prospect is that of the all-knowing rationality of a fair and impartial electronic decision maker. The "oraclisation" of the fourth-party role could work well in the construction sector through the medium of blockchain and smart contracts provided access to the data could be regularised and structured in a way that an AI could process. Further, the contextualisation of the truly important points of disagreement in a dispute and comprehending the subtexts and assumptions behind each of those points will remain a challenge for AI in the foreseeable future.

These and other issues were considered by the UK Law Tech Delivery Panel's Jurisdiction Taskforce[12] in addressing both the shape of things to come and some of the current issues preventing development. In short, the statement is a plea to the legal and technology communities to work together in building dispute-resolution mechanisms into smart contracts.

> *My hope is that English law and our UK jurisdictions will be able to provide state-of-the-art dispute resolution mechanisms specifically tailored to inclusion in smart contracts.*

The statement recognised that the soon to be 3 trillion smart financial services contracts globally every year would likely involve some significant disputes.

Lord Justice Vos foresees these developments:

1 Small disputes will be resolved by ODR processes including mediation.
2 For larger disputes, there will be a need for a carefully restructured approach to the way we deal with evidence, use AI to resolve the dispute itself. Meeting the expectations of the new generations of business disputants is key here.
3 The courts do not have capacity to deal with disputes in all sectors, and we need to make more structured use of ombudsmen, mediation and early neutral evaluation to provide dispute resolution that is better tailored to each particular type of dispute.
4 The way we resolve smart contracts disputes will be critical to the rule of law in the future. Courts will need to ensure they can remain relevant to dispute resolution in legal coded contracts on the blockchain or its equivalent.

This approach is a reminder that civil justice systems exist to offer a service, albeit a service that no one hopes they will need to use. The statement is helpful in removing pre-conceptions on what the judiciary stands for and should lead to a healthy dialogue. Certainly, the envisaged removal of the time-honoured stages of litigation – pleadings, documents and examination of witnesses – would go a long way in reducing time and costs provided the forensic checking of records and externalities achieved the same outcomes. The dispute-resolution procedures of the future will need to be light touch, economical and accessible. Notwithstanding this, the availability of independent judicial dispute resolution is critical if the technologies that underpin that smart contracts are to secure the confidence of mainstream investors.

Lord Justice Vos also points out the coder's reluctance to engage with lawyers and the legal system. There is also a strong desire amongst crypto-assets and smart algorithms for disintermediation. The important thing will be to devise an approach that will bring the technological community on board. A menu of ODR processes is envisaged that allow the parties to choose a process that meets their expectations in terms of cost and speed of outcome.

11.4 Dispute resolution, the steps in-between

ODR is often referred to as a form of ADR that takes advantage of the speed and convenience of online platforms. ODR has been used extensively for e-commerce particularly in relation to complaints and disputes that are cross-border, low-value, high-volume and take place between online users.

ODR came about as a response to the "wild west" of the explosion of the internet in the 1990s. It was recognised that trust had to be built in the e-commerce platforms and this involved the institutionalisation of avenues for addressing and preventing disputes.

The most recognisable amongst these ODR platforms are eBay and PayPal that employ a tiered ODR process where parties first attempt to settle their disputes using assisted negotiation software. If this fails then the claim escalates. PayPal goes a stage further and freezes the money involved in the transaction in dispute in order to ensure the enforceability of the final decision. A similar approach would appear logical in a smart contract scenario. PayPal resolves over 60 million disputes a year in this manner.[13]

The growth of ODR has been relatively slow outside of a few specific online market places. The reasons for this include lack of incentive for traders using external ODR and the perception that these are biased entities. This mistrust is stronger where the ODR provider is a private entity. Another issue is on the low-value nature of the disputes and the appropriate use of procedures. The enforceability of resolutions is a further difficult issue.

The public sector has had more success in designing and implementing an ODR platform. Personal injury claims in England and Wales represent a considerable proportion of the litigation occurring in the courts and streamlining these processes has long been high in the judicial review agenda. The Claims Portal was established in 2010 and is an electronic platform for low-value personal injury claims. Claims with a value between £1,000 and £25,000 involving road traffic accidents, employer's liability (accident and disease) and public liability accidents have to started online using this platform. In 2017, over 827,000 claims were submitted via the Claims Portal and over 192,000 were settled.[14]

Claimants using the portal can issue as litigants in person or can employ solicitors. Solicitors' recoverable costs are strictly controlled and subject to stringent limits. The role of solicitors here is subject to the disintermediation movement discernible amongst professionals more widely. The judicial reform groups would rather remove solicitors from the picture completely on low-value claims and a claimant-operated ODR is in development. However, there is a balance to be struck between accommodating litigants in person and the right of a defendant to know the case they have to answer and to see a properly articulated and drafted pleading with evidence. The point is arrived at where the residual value add of a professional (here a solicitor) should not be eroded further without proper regard to the consequences.

The choice for ODR is to either inject data about the physical world into a digital ledger – either place trust in traditional existing mechanisms to arbitrate disputes, or try to create a resolution mechanism with the system itself.

In either case, the immediacy of the dispute resolution is in everyone's interest leading to a speedy resolution where arbitration can be seen as a service (such as Ebay's Resolution Centre) and not necessarily a relationship ending or a costly affair. There is a new driver towards successful mediation of the dispute – no party wants a negative review of their status as buyer or seller as this might adversely reflect on their business or trading status.

One of the issues of using traditional courts and legal systems is the issue of jurisdiction. The discussion on where a contract is formed when the different members of a distributed ledger network might be based in different countries with different expectations of legal process and different interpretation of agreements is a challenge. More importantly, traditional dispute-resolution mechanisms are precisely the types of systems that are currently deemed as inefficient and expensive, and therefore often what the adoption of smart contracts is intended to minimise. For this reason, some business are exploring new ways of resolving dispute using the distributed nature of the systems themselves. Systems such as Kleros, Juris and Sagewise that rely on members of the network to act as arbiters have the advantage of bypassing the need to trust a central administrator to define what counts as truth within the ledger, thereby avoiding the costs associated with traditional mechanisms.

The consensus governance mechanisms and distributed pools of arbiters could be used to resolve disputes and how economic incentives might be set up to reward just arbitration. This effective "wisdom of crowds" approach to dispute resolution appears to have a good deal to commend it. Another approach is to use a Human API – application programming interface.– Amazon Mechanical Turk (MTurk) is a crowd-sourcing website for businesses (known as Requesters) to hire remotely located "crowdworkers" to perform discrete on-demand tasks that computers are currently unable to do. It is operated under Amazon Web Services[15] Employers post jobs known as *Human Intelligence Tasks* (HITs), such as identifying specific content in an image or video, writing product descriptions or answering questions, among others. Workers, colloquially known as *Turkers* or *crowdworkers*, browse among existing jobs and complete them in exchange for a rate set by the employer. The nature of jobs might include anything from conducting simple data validation and research to more subjective tasks like survey participation and content moderation. One stated aim of the project is to accelerate machine learning development.

Future generations of business disputants are likely to want their remedies in real time. Real-time resolution lies within the gift of digitisation. This accords with an earlier point about disputants being prepared to live with a "rough and ready" adjudication decision rather than waiting for the greater scrutiny of a litigated outcome. How much more acceptable then is an immediate resolution of issues without any of the usual anxiety and lack of control over the spiralling time and cost of traditional resolution procedures.

The role of collaboration is already delivering some real-time results in terms of reducing disputes. Collaborative risk management now regularly involves additional reviews and enquiries through which team members challenge their own and each other's role assumptions when there is still time to take mitigating actions without causing project delay. The incidence of disputes can be reduced further by the technologies outlined in this chapter and the challenge is there for the latest versions of the standard form contracts to engage in these innovations where possible in the complimentary areas of dispute avoidance and mitigation.

Notes

1. Speech delivered by Lord Donaldson in 1986, cited in Chern, C. (2011) *Chern on Dispute Boards: Practice and Procedure*. Wiley-Blackwell, Oxford.
2. Arcadis Global Construction Disputes Report 2019, available at: https://www.arcadis.com/en/united-kingdom/our-perspectives/2019/june/global-construction-disputes-report-2019/ last accessed 13 July 2020.
3. [2017] EWCA Civ 117.
4. Macob Civil Engineering Ltd v Morrison construction Ltd (1999) EWHC Technology 254 Lord Justice Dyson.
5. (2005) EWCA Viv 1358.
6. FIDIC Conditions of Contract for (i) Construction (Red Book), (ii) Plant and Design Build (Yellow Book) (iii) EPC/ Turnkey Projects (Silver Book).
7. The New York Convention on the Recognition and Enforcement of Foreign Arbitral Awards (1958) – enforcement of foreign arbitral awards on a reciprocal basis.
8. Charted Institute of Arbitrators (2011) *Costs of International Arbitration Survey* available at: www.ciarb.org.
9. https://www.npr.org/sections/parallels/2014/11/05/359830235/brazil-the-land-of-many-lawyers-and-very-slow-justice?t=1593940753387.
10. https://timesofindia.indiatimes.com/blogs/toi-edit-page/what-causes-judicial-delay-judgments-diluting-timeframes-in-code-of-civil-procedure-worsen-the-problem-of-adjournments/.
11. Rabinovich-Einy, O., & Katsh, E. (2019) *Blockchain and the Inevitability of Disputes: The Role for Online Dispute Resolution*, Journal on Dispute Resolution, 47.
12. Lawtech Delivery Panel (2019) *Statement on Cryptoassets and Smart Contracts*, available at: https://technation.io/about-us/lawtech-panel last accessed 21 May 2020.
13. CSLS Oxford, 28 October 2011, What should the ideal ODR system for e-commerce consumers look like? The Hidden World of Consumer ADR: Redress and Behaviour.
14. Wallis, T. (2017) *Legislative and Regulatory Moves in England and Wales Impacting on the Future of ODR*, International Journal of Online Dispute Resolution, 4(2).
15. Amazon Mturk, available at: https://www.mturk.com/ last accessed 13 July 2020.

12 Conclusions and next steps

12.1 Conclusions

Albert Einstein is credited with saying *"the definition of insanity is doing the same thing over and over again, but expecting different results."*[1] Technology provides a different way of doing things and therefore a brilliant opportunity. The fundamental choice is whether to stand aside or to become involved in the movement. It is tempting for those of us about to enter our sixth decade to wait for the next generation of digital natives to drive change. However, linking the past, present and future is essential in order to ensure a smooth transition and to give the future initiatives the best chance of success. The built environment has excellence in its structure and people represented by good practice but suffers from a poor reputation as being financially uncertain, slow to change, adversarial, untrusting and unfair to subcontractors. It is a deeply vexing question as to how such a massively important player on the world's scene perpetuates its shortcomings. The answer to the problems would appear to be to embrace technology.

During Caesar's conquest of Gaul, it became necessary to secure the eastern border of the new provinces against marauding Germanic tribes. The tribes felt safe on the eastern side of the Rhine river (Figure 12.1), trusting the river as a natural border which offered cover from retaliatory attack after their opportunistic raids into the province. Caesar decided to confront them. While he could have crossed the river by boats which the Ubians had offered to provide, he decided to build a bridge, thus demonstrating Rome's ability to bring the fight at any time to the Germanic tribes; and additionally, as he indicated in his *Commentary on the Gallic War*, this approach conformed more to his own dignity and style, than to anything else.

The construction of this bridge showed that Julius Caesar, and Rome, could go anywhere, if only for a few days. Since he had over 40,000 soldiers at his disposal, they built the first bridge in only 10 days using local lumber. He crossed with his troops over to the eastern site and burned some villages. However, the tribes had come together and were prepared to meet Caesar's army in battle; but when Caesar heard of this, he quickly left the region taking down the bridge behind him.

Taking down the bridge was a secure way of making sure one is not pursued. Another reason could be that to leave a marvel of engineering in place and to give away the construction technology was not in the Roman's best interests. It is

Figure 12.1 Caesar's bridge over the Rhine

stretching the analogy to suggest that the construction industry represent tribes-people requiring subjugation by a superior enhanced force. However, in there is the nucleus of the point being that the construction industry must indicate that we are ready for the technological solutions and will not continue to shun them for short-sighted reasons.

The availability of technology-led innovation should not be considered in a vacuum. We are able to show how robots – whether humanoid or flying – can build a brick wall. This does not though equate to on-site reality until a host of other obstacles have also been overcome, foremost of which is the legal under-pinning. The clients themselves have to want to invest in the changes required and to accept the case for their adoption. As things stand, the client's desire for innovation is limited, their procurement decisions depend on transactional models and behavioural norms which promote cost, not value. Re-education and culture shift are required. Happily, there are signs of this new approach in the public sector around the concept of social value.

The central development and enabler of all the technologies discussed in this book is the internet of things – ubiquitous and pervasive computing which embeds processors, sensors and internet connectivity into physical objects. Tiny connected computers planted inside everyday things open the pathway to the related developments chronicled here. The ability is to bring people together as never before in the built environment. Similarly, the ability to digitally automate a process whether it be in design, costing, transacting, or robotic manufacturing or assembly, requires a newly defined interface between human values and machine process. This is the need for a new digital skills agenda and sophistication.

The role of BIM has been discussed as part of this accelerating and growing movement towards digital sophistication. The issue is now how well we manage the transition of BIM's maturity and how well it is used to address societal and commercial interests. BIM can, at times, appear impenetrable and unworkable especially when the guidance notes are changed and a new set of instructions is given. Nevertheless, the growing pains of an emergent technology require patience and dedication from those whom it would serve, as well as a good degree of perseverance.

Looked at positively, the technology journey in construction can be likened to the health and safety emphasis of the 1990s and the sustainability focus of the 2000s. Both approaches are now so embedded in the approach to construction as to make it inconceivable that they would not exist. BIM and its related technologies are part way towards the same level of adoptions and penetration into the collective psyche of the construction industry.

Questions have been raised around whether a more intuitive approach to digital innovation could, in time, subsume BIM into it. The starting point for this discussion to strip back transactions to their core obligations of selling and buying services and/or goods. Doubts can also be raised about the truly transformative nature of what has been proposed. The fifth element of the Digital Twin maturity model raised a question of the purpose of seeking to replace humans as the desired end goal. Further, doubts can be raised as to whether any such thing is achievable. At one level, the oracle role discussed looks a lot like the traditional Architect/Contract Administrator role, albeit a disaggregated one with a mystical name. Unbundling a professional's role and devaluing their professional judgment and input could be seen by some as a retrograde step. A reappraisal of whether this is definitely what society wants from its professionals ought to be instigated before it is too late.

This quandary may never require resolution if the construction industry reverts to type and does not go through with the changes envisaged. For many, any discussion on the fourth industrial revolution underpinned by cyber-physical smart is woefully premature in an industry yet to achieve Industry 3.0 status in relation to a proper application of data-processing technology.

The book has suggested that current debates around such issues as good faith and blame-free construction in the collaborative movement will also turn out to be academic in a technologically enhanced future. Any approach requiring the exclusion or limitation of reasonable legal rights and obligations ought to be viewed with scepticism. Legal issues remain in relation to those initiatives that are established and are in the course of becoming established. Multi-party contracts present issues around the ability of each participant to bring and defend legal actions where liability needs to be established. The protection of intellectual property rights and business intelligence is a key concern for the supply side of the industry moving forward.

Another key conclusion from the work is to recognise the limitations in our own relationship with technology. The artificial intelligence fallacy needs to be borne in mind. Our expectation that the only way to develop systems that perform tasks at the level of experts or higher is to replicate the thinking processes of human specialists. This anthropocentric view of smart systems is limiting. Just as with the Alpha Go programme, systems of today are outperforming human experts, not by copying high-performing people but by exploiting the distinctive capabilities of new technologies, such as massive data-storage capacity and brute force processing.

12.2 Next steps

The least likely future of all is that nothing much will change. Even if there are no advances in technology in the next decade as fundamental at the Personal computer, the World Wide Web and social media, if we follow existing and emerging technologies to their probable and much greater exploitation then this alone takes us into a very different world. The least likely future for our technologies is that our systems will stay the same as they are today. The power of data is seemingly infinite. Gigabytes give way to terabytes, which give way to exabytes. This exponential growth in information technology and our increasingly capable machines, increasing pervasive devices and increasingly connected human beings build a compelling case to silence the doubters.

That is not to say that these increasing capabilities should be squandered. The current trends to throw everything into the cloud without bothering to delete anything appears wasteful. This results in more data centres and more capable processers, which is, consequently, one of the biggest growth areas for the built environment. Other scientists seek to exploit the laws of physics to their nth degree by introducing nano-materials and encoding data on sub-atomic particles. Ultimately, sustainable approaches to data use are likely to be required.

The pace of change is of similar great importance. New exponents appear impatient for the upheaval in the short-and-medium term. Older heads and more established providers of services anticipate a more leisurely and civilised transition. Black swan events, which seem to occur with a greater frequency, will have the final say in the pace of change. It is often in response to these economic or tragic events, such as the Grenfell fire that lead to the clamour for new forms of contract and procurement routes being developed. The key here is to change the structure of the contractual networks adopted. The limitations of the linear approach have been reviewed as well as the benefits of the hub-and-spoke model with its built in compatibility with BIM as the control centre. Actual overlays planned and the dashboard of completion demonstrates progress against pre-decided metrics and visualisations for the project manager whilst retaining control over payment released against earned value. The providers of standard form contracts in the United Kingdom stand at a crossroads of accepting this type of multi-party contract as the way forward. For the most part, construction contracts remaining resolutely linear seem set to perpetuate poor practices in terms of payment abuse and unfair risk allocation.

The hub-and-spoke model itself becomes redundant when the data-processing capability takes over. The stigmergic approach becomes preferable from a management point of view given that it facilitates automation and optimisation.

However great the data-processing capability becomes and we have seen that this power grows exponentially, AI and fully automated smart contracts will continue to struggle with the issue of context for some time to come. Every building site is different and the constraints and factors coming together, unless taken out of the equation, for instance in a factory setting, will require the human input in the foreseeable future. Similarly with predicting the use of AI for dispute resolution – will the robot judge have understood the real context to the dispute and the subtly important subtexts? Perhaps, once enough cases have been studied then this machine learning could evolve.

The most-likely scenario for the smart contract agenda appears to be the stack approach, where automated provisions of contract reside side-by-side with traditional obligations inside a smart contract appear to be the next logical development. The risk in this field is to ignore the importance of continuity for the existing provisions to the future ones envisaged. A tendency exists to seek to simplify the law amongst those flummoxed by its vagaries it. An example is in the integrated project insurance movement within construction that seeks to provide blame free construction. Parties to an IPI project may only sue each other in the event of "wilful default". This represents a departure from a whole body of law that has been created around negligence and the need for accountability in the fulfilment of professional roles. This might be a fantastic new initiative to cut through complexity. Or, it might represent a rush to gloss over the nuanced reality, which the law exists to decipher. The legal community needs to be part of the debate seeking to develop a formal contracting language capable of dealing with these subleties.

The requirement of blending the different stances together into a harmonious approach is the ultimate message here. Pursuing full automation without a rationale for it leads to mixed messages and second guesses about the reason and direction for travel. Better results are available through pursuing the development of collaborative procurement and BIM alongside the development of smart contracts, Internet of Things and Artificial Intelligence. Technology represents a viable solution to the built environment and wider societal problems and its ability to inform and support human interactions in helping us focus on the optimum approaches is its key usefulness.

By putting these measures together, it is possible to emerge on the other side of the portal. The final sentiment on which to leave this work is the reminder that complexity is not, of itself, insurmountable.

Note

1. Einstein, A. (Republished 2015) *Letters to Solovine: 1906–1955*. Citadel Press, New York.

Index

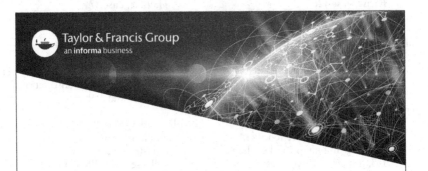

Printed in the United States
By Bookmasters